PCBARTIST 2 WITH LTSPICE DESIGNERS GUIDE

by David Hunt

Schematic Capture, Simulation and
Layout of PCB with Free Software

Table of Contents

Introduction..**8**

    Aaron Every Step Books...............................8
    PCB Artist...8
    LTSpice...8
    Texas Instruments..9
    Aanoit Technical Services............................9
    David Hunt...9
    Advanced Circuits.......................................10
    Apache Open Office......................................10
    Disclaimer..11
    Acknowledgment...11

PCB Development Work Flow.................**12**

 Design Flow..**15**

Preparation...**18**

    Functional description.................................20
    Interconnection..20
    Interaction..21
    Power requirements....................................21
    Component selection...................................21
    Mechanical requirements...........................22
    Environmental requirements......................22

Creating a new Project.............................**23**

 The New PCB Board Wizard.....................**26**

    Board Size Selection..................................28

Design Requirement Selection..............**30**

    Layers Selection...33

Board Parameters.....................................**35**

Additional Requirements..**36**

    Board quantity and turnaround time........................37

PCB Design Name..**38**

Preferences..**40**

    Display..42

Interaction...**44**

    General Interaction...45

    Schematics Interaction...46

    PCB Interaction..46

    Adding PCB Tracks...46

Cross Probe..**48**

Schematic Capture...**49**

Connections..**49**

Components..**52**

The PCB Artist Libraries...**53**

Creating Surface Mount Schematic Symbols...............**56**

The Library Manager..**57**

Creating A New Library...**58**

    Schematic Symbols..59

Schematic Symbol Wizard..**63**

    Left and right pin placement.................................67

Creating Through Hole Schematic Symbols................**71**

PCB Symbols..**81**

Creating Surface Mount PCB Symbols.........................**84**

Creating a New PCB Symbol Library............................**87**

    PCB Symbols...88

    Solder Mask..90

Creating Thru-hole PCB Symbols.................................**94**

Technology: **95**
    Units: 96
    Precision: 96
Footprint **96**
Pads **97**
    Pad Counts 98
    Pad Style 98
    Measurements 98
    Silkscreen 100
    Layer 100
    Position Silkscreen 100
    Include Marks 101
    Placement Outline 101
Finish **102**
    Footprint Name: 102
    Save the footprint to the library: 102
    Edit the footprint now: 102
Editing the PCB symbol **105**
The Silkscreen **109**
Creating New Components **111**

Wizard **111**
Component Details **112**
Schematic Symbol Selection **113**
PCB Symbol Selection **114**
Pin Assignment **116**
Library Enhancements **117**
Schematic Reports **119**

    Dangling Tracks 119
    Design Status Report 121
    Generic Netlist 122

Unconnected Pins...126

Entering the Schematic.................................128

Placing a Component.................................129

Dual Selection...132

Setting the Grid.......................................136

Adding Components.................................136

Wiring components..................................139

Bill of Materials...147

Viewing Component Values.....................148

Component List:......................................153

Editing the PCB – Placing components.........156

Error Correction......................................167

Editing the PCB – Routing Components.........171

Pad Styles...178

Changing Net Classes.........................179

Updating PCB from Schematic...............181

Autorouting...183

Post Autoroute Cleanup..........................190

But Wait!...192

Let's Spice It UP!.....................................197

SPICE History...197

Linear Technology..................................198

LTSpice IV..198

How It Works..199

Limitations of SPICE..............................200

Your First Simulation..............................201

Creating a schematic in LTSpice...............201

Creating a SPICE schematic in PCB Artist......206

The LTSpice Display..**213**

SPICE Measurements..**215**

Updating Version 1.5 components for Spice..................**219**

    Updating Standard Devices **219**

        The Resistor 219

        The Capacitor..222

    Updating an OpAmp **228**

    Creating a Passive SPICE X model...............................**230**

        Adding the potentiometer to the circuit 237

    Creating an Active Component Model..........................**240**

        Component Type..241

        Component Details...241

        Schematic Symbol...242

        PCB Symbol...242

        Assign Pins...242

    The Hack..**245**

The OPAy322.lib library...**249**

Spice Suffix Nomenclature...**257**

Typical schematic symbols..**258**

Audio Mixer Product Design Requirements...................**259**

SO-8 Package..**261**

P140KV1 Through Hole Package..................................**262**

Schematic Drawings...**263**

Introduction

Aaron Every Step Books

The Aaron Every Step series of technical books is written to demonstrate every step necessary to use the product being described. Unlike books that simply rehash the menu items, this series sets a goal and describes every step necessary to achieve that goal.

The goal of this book is to serve as a supplement to existing PCB Artist documentation. This book supplements the on-line help and tutorials and gives application specific information regarding the application of this software to a real world project for the 1st time user as well as the seasoned PCB designer.

The instructions and tutorials are excellent and should be used with this book, especially during routing and placement. In fact, the tutorial should precede reading the book.

PCB Artist

PCB Artist is professional grade printed circuit software distributed freely by Advanced Circuits. The software is fully functional in its free form and there is no cost for all capabilities and options.

The software contains schematic capture, printed circuit board layout and autorouter capabilities. The software can be downloaded at:

www.pcbartist.com

LTSpice

Linear Technology provides state of the art components at a reasonable price to the electronics industry. In my opinion they are the analog masters with a broad line of reasonably price high quality integrated circuits.

They also provide and support LTSpice, a free, high quality derivative of Berkeley Spice. LTSpice is supported by PCB Artist with a nifty one-click interface.

Although they don't offer a money back guarantee their software does everything they claim and more. I highly recommend its use.

http://www.linear.com/designtools/software/#LTSpice

Texas Instruments

One of the original players in the integrated circuit market TI has proved they belong in Texas where everything is bigger, and better. Like Linear Technology they provide state of the art components that work as advertised. I am a circuit designer and will purchase from either company without hesitation.

I elected to use a TI device because I needed to show how to use the library component of another vendor. I also use TI's SPICE software TINA and find it quite acceptable.

www.ti.com

Aanoit Technical Services

Formed as Bama Instrument in 1992 and renamed in 2005 Aanoit is pronounced an-wah which is derived from a Finnish word for provider. Aanoit provides technical services to industry from its location just outside the Oak Ridge National Laboratory in Oak Ridge, Tennessee.

www.Aanoit.com

David Hunt

Mr. Hunt has been an engineer since the 1970's and hardware and software contract engineer since 1995. As a consultant he has spent much of his time traveling the United States developing electronic products from concept to delivery in medical, aerospace, consumer, military and other markets. This wide base of experience has given the author insight to a large number of software tools and methodologies.

Having worked in many start-up companies the author has seen massive amounts spent on software tools with capabilities much less impressive than the PCB Artist tool set.

This tool set has been selected because it is complete, stable and capable.

www.DavidHunt.net

Advanced Circuits

When PCB Artist software is used to generate a printed circuit board file it is sent to Advanced Circuits via the Internet where it may be used to generate a printed circuit board. The printed circuit boards generated are of the highest quality and may be generated for a reasonable price. Advanced Circuits is a United States company with many manufacturing facilities including Colorado, Arizona and Minnesota.

Without qualification, the author recommends Advanced Circuits to any potential user for prototype boards and for production boards due to their quality product and on-time and on-price performance.

There are some very special programs for college students.

www.4PCB.com

Apache Open Office

How appropriate that a book about free printed circuit artwork be written using free word processing software. Open Office is capable of almost anything the 'bug guys' (not a misspelling) can do – just better. Yes, I have a copy or two of Word, it just isn't as capable for a book writer. However it insists on capitalizing the first character of web addresses.

Www.OpenOffice.org

Disclaimer

The author has no affiliation with Advanced Circuits or PCB Artist except as a satisfied customer who has generated a number of projects and the resulting printed circuit boards using this software.

I express a number of opinions in this book, it's my book. Unless stated as fact, all statements expressed are opinions.

Acknowledgment

To Marie, my enabler.

PCB Development Work Flow

Development of any printed circuit artwork should begin only after all the items in the Preparation Chapter have been gathered. This entire process should be considered as interactive. The design process may or may not be enhanced with increased interaction, primarily depending on the individuals involved.

For well-defined tasks, interaction only expends valuable resources and work is duplicated. However, some interaction is necessary. For example, a machinist will have skills not possessed by the circuit designer and their tasks must be coordinated .

The most severe and devastating American Manufacturing curse is "I'll know it when I see it." Managers that recant that phrase almost never know 'it' when they see 'it'. They spend their time waiting for someone to show 'it' to them. They will quickly tell you that "It is your job to present me with alternatives."

An alternative managerial process is to create a written set of requirements so that every person on the team can "know it when they see it." A statement of requirements allows the individuals of a team to perform their individual function and creating a well-defined and operational product. This was the germinating concept of ISO-9000 philosophies.

As my proof of concept I present Mr. Hewlett and Mr. Packard[1]. They even shared some of their requirements to the rest of the engineering community. Unfortunately, those that

---

[1]One story about the pair is that they tossed a coin to determine the order of the company name. Mr. Packard won the toss and decided to call it Hewlett-Packard. Imagine that happening with a modern day middle manager!

would have "known it when they saw it" apparently didn't look at it. And we lost the electronics industry to overseas providers.

The requirements consist of all the information available about the end product.

The scope of this book is the creation of printed circuit boards utilizing PCB Artist software.

Requirements for a finished printed circuit board design include, as a minimum:

- Circuit design – including schematic, bill of materials, functional requirements. Circuit design is an iterative process that is influenced by cost, manufacturability, component availability, available resources and other considerations. Especially for new products, the design should change as product knowledge evolves. This is the role of the circuit designer.

- Schematic Capture – The events of schematic capture include creation of models, schematic entry, interconnection and finally review by the circuit designer. Schematic entry can be the role of either the circuit designer or the printed circuit designer.

- PCB Layout – PCB layout is the creation of artwork that may be used by a printed circuit manufacturer to create a board. Usually, most of the manufacturing information is also created during the PCB Layout process. Again the final PCB layout is reviewed by the circuit designer.

- Component library – Schematic design and PCB Layout require the use of symbols, also referred to as models, to be developed for both immediate visual representation to aid in the design process and to be used by the software to generate the reports and information required to generate and manufacture a printed circuit board.

Many PCB capture/layout products separate the schematic capture and PCB layout portions of the design utilizing separate schematic capture and PCB layout libraries. PCB Artist is unusual in that the component libraries that include both the schematic model and the PCB symbol. To create a component one selects or creates a schematic symbol then selects or creates a PCB symbol and the combination is represented as a component. The advantage of this methodology is that developing both models simultaneously results in a reduction of total development time.

For example, if a microprocessor is available in a number of packages a separate component model can be created for each of those packages. There might be more pins in one package than another. There might be a different pin out (assignment of pin number vs function.) The same schematic symbol can be used for both packages and a different PCB model selected. Instead of having to develop a new schematic package, the pin mapping is changed at the component level. The component model greatly simplifies this situation. I typically use generic models in the schematic capture symbol and PCB symbol and create a component that utilizes the complete component model name, model, ordering information and, now, in Version 2, SPICE model information.

The result is libraries that are and remain quite constant during the life of a product. If a component must be changed those changes must be incorporated into the component library and not into the schematic or PCB models. The component model can have information added that is specific to that component such as its ordering information. This book will explain how to include accurate bill of material information in the Component model and this document will explain that process.

Design Flow

Each step in the design flow drawing below corresponds to traditional areas of responsibility. Printed Circuit requirements should be an integral portion of the total design flow. Even if the total project is being designed by one person it is advantageous to document the individual stages in case the person obtains help, say due to a change of schedule, or wants continuity as in the case of a vacation or other planned or incidental interruption.

New product development should be an iterative process. That is why capture of information for individual steps in the process is important. As the design matures it sometimes becomes obvious that preferred alternatives exist and the project modified to take advantage of the increased information.

A sample requirements document for the Mixer we are designing is provided as an attachment. In larger companies the product design requirements are often created by marketing and management. The electronic design requirements are generated to match the product design requirements. The printed circuit design requirements are then generated from the electronic design requirements.

The following flow diagram must also be considered as fluid. For example, I design circuits with the schematic capture tool that I'm using at the time. This requires some symbol development before and during schematic capture, swapping the order of those elements in the process flow. The flow below might be for a paper napkin circuit design that is subsequently embellished by a printed circuit designer.

The Bill of Materials must be generated early because all the schematic capture and printed circuit layout must be developed to EXACTLY the device(s) that are to be utilized in the specific design.

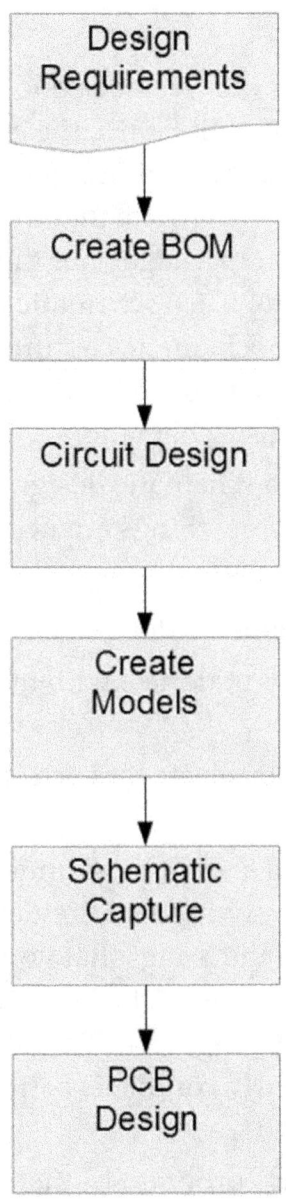

Preparation

There are a number of roles in developing a project. When developing a printed circuit board, the essential roles include the circuit designer, mechanical designer and schematic/PCB designer. There is a good chance that the circuit designer will use a schematic capture package, but there is a chance that the circuit designer will not use a schematic capture package or will use an alternative schematic capture package. In either case the PCB designer will have to convert the schematic provided to a PCB Artist schematic.

To properly create a schematic design and/or a printed circuit board design a few things are necessary.

- Circuit design
- Written requirements document
- Complete bill of materials with manufacturer's part numbers
- Component data sheets for each item on the bill of materials

None of these items should be assumed or implied. Each requirement should be written and reviewed by the team. The requirements provide the things that you must know as a schematic and PCB designer. The items above must be provided by the design engineer. This attitude may seem confrontational, but early conflict is better than conflict after the budget is expended.

A schematic capture program is often used in design even before a functional description is possible. The schematic brings most of the electronic design pieces together in one place and gives the designer an excellent overall conceptual view of the project. I personally use the schematic capture program as the primary design visualization tool.

However, this book is written to provide a methodology of creating schematics and artwork. For the purposes of this description the information might be transferred to the practitioner in the form of a pencil drawing on a napkin.

In each case the act of design is usually an interactive process with each step requiring a reevaluation of the previous step in the design process.

The list below is not being presented as an exhaustive checklist. Each development project is different. It is presented to demonstrate a set of requirements necessary for use of this set of tools.

1. Functional description
2. Interconnection diagram
3. Power requirements
4. Interaction
5. Component selection
6. Mechanical requirements
7. Environmental requirements

Functional description

The assumption is that you have a design in hand and need to generate artwork including the schematic and/or printed circuit artwork. Many engineers will work with the schematic software directly at the conceptual stage of design making this assumption invalid. In this case this portion of the discussion can be considered a PCB preparation stage.

The function description is a statement of all the functions a design must accomplish. This is necessary for the schematic and PCB portion of a design because many characteristics are determined by function. For example, a transistor may be selected because of gain and bandwidth considerations or it may be selected for thermal characteristics. If gain-bandwidth considerations are important, low capacitance techniques may dominate such as minimal line width and short traces.

However, if thermal considerations are dominate, then allowances for heat sinks and extra line width for increased current, extra spacing for high voltage isolation and extra line length to facilitate heat sinks may be most important.

Even the most trivial known fact is significant and should be included in the functional description. An example is the color of components when analyzing thermal circuits.

Interconnection

An electronic product will probably have external connections of some type. Those connections might be RF based (such as WiFi connections), capacitive based (such as tablet touch screens), conductor based (such as USB cables) or audibly based (such as radios.) Each interconnection technology has its own characteristics and challenges. Most have very specific circuit and PCB requirements.

In today's world of miniaturization a product's connectors may occupy more product volume and PCB space than all other components combined. This, in-turn, directly equates to product costs including but not limited to component costs, shipping costs, storage costs, inventory costs and so forth.

Interaction

Some interaction is obvious such as a component that is too tall to fit in the final case. Other interactions are less obvious. An example is a component that generates enough heat to cause the failure of another component. Connector placement and alignment should be determined early in the design process.

Power requirements

Although it may be considered by some as a "no-brainer", power generation and availability have caused a large number of problems in a large number of projects. Power requirements should be determined early in the design process.

Component selection

Some components are critical to a project and a large part of the design process is dedicated to selection and proper utilization of a key component. Some components are taken for granted, such as the resistor. It is always a good idea to occasionally review each and every component in a project to determine if all requirements are being met.

This is one of the primary reasons for periodic critical design reviews. In this case critical refers to the attitude of the reviewers. A good critical reviewer will make no assumptions nor take any consideration for granted when reviewing the project. A good friend will find all the flaws.

Mechanical requirements

There are many variations to these requirements and they change from project to project. Some of the recurring questions include:

After a printed circuit board has been generated how will it be mounted?

Will a heat sink be required?

Are the board dimensions such that it requires additional support?

How will the end user interact with the product?

Is there a need for human access to switches, screens or other product features?

Environmental requirements

This refers to environmental considerations such as temperature, humidity, radio frequency emissions, radio frequency susceptibility and external heat sources. Temperature extremes may require heaters (as is the case with LCD displays) or coolers (as is the case with power amplifiers or computer power supplies.)

Humidity may require additional processing such as a sealant or encapsulation.

Radio frequency susceptibility (RFI) may require a metal enclosure.

External heat sources might require a fan (and its power source.)

Creating a new Project

PCB Artist software is truly an integrated package with schematic capture, PCB design and SPICE emulation. The Project is the glue that ties all these packages together.

If a project has been created there will be a Project in the main menu bar.

A quick note regarding working with printed circuit artwork software. All PCB work in either the layout editors or library editors is from the top side of the board. If you have a board placed in front of you with the components facing up then looking down on the board is the view you will use for all work with this package. This is the industry standard.

There is a chance that the schematic capture will be performed by another person and the PCB design performed by someone else. In this case the information might be transferred by a single schematic file. In this case the schematic would be existing before the Project. To view the project type Alt-P and select the file with the .prj extention.

The project page gives an overview of the current project. We have a schematic (Mixer.sch) but no PCB. This section is not to add components or to wire the PCB it is to set up the correct environment for creating PCB artwork. I suggest that you not bypass this chapter it is an essential part of the design process.

To create a new printed circuit board (PCB) select File-> New.

You will get the New Design window select the New Project radio button and press OK.

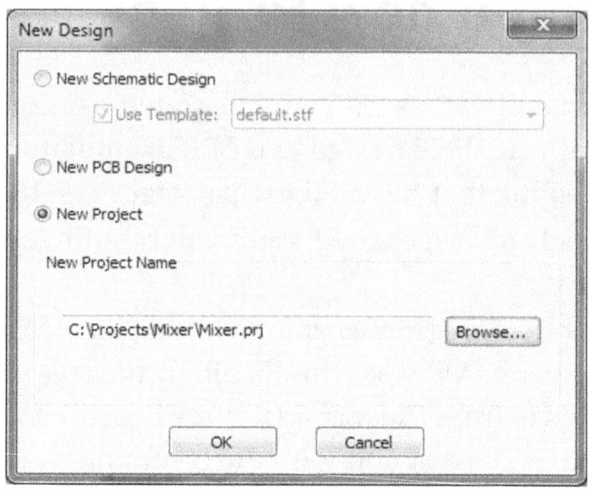

This screen is rather insistent that you first Browse for the project directory utilizing the pull-down browse button. Then you can type in your project name. Mixer for this project.

Select OK.

You should see something similar to the following image. You may notice that I 'scrunch' the total screen down so that I can have good resolution for the important information, such as the Mixer.prj Window in this screen.

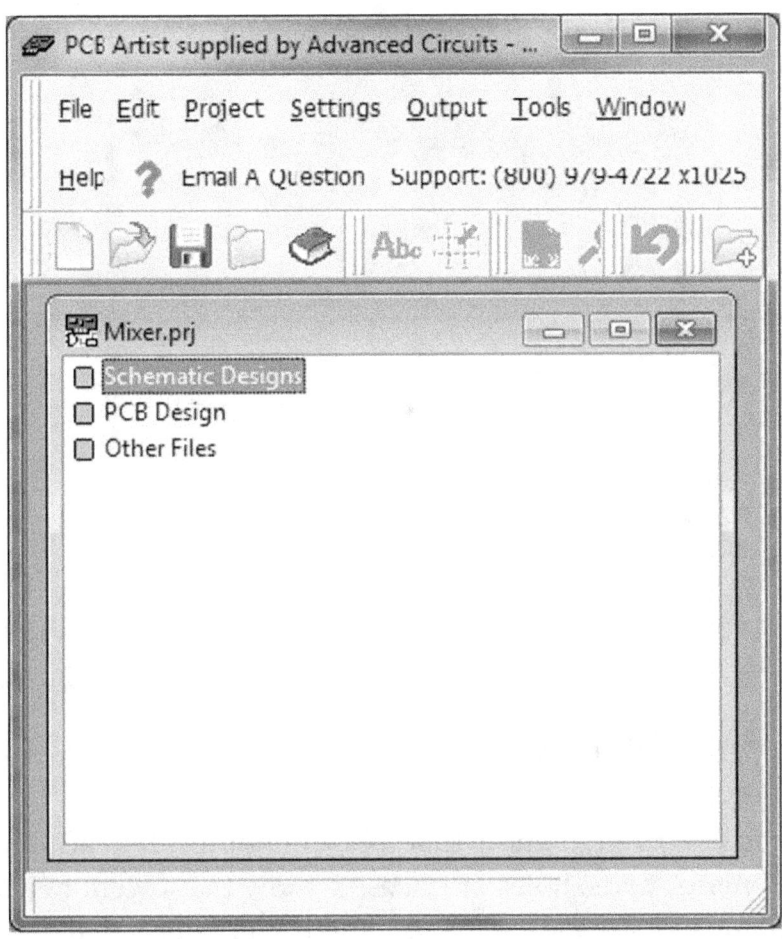

Select Alt-F then N and select the New PCB Design push

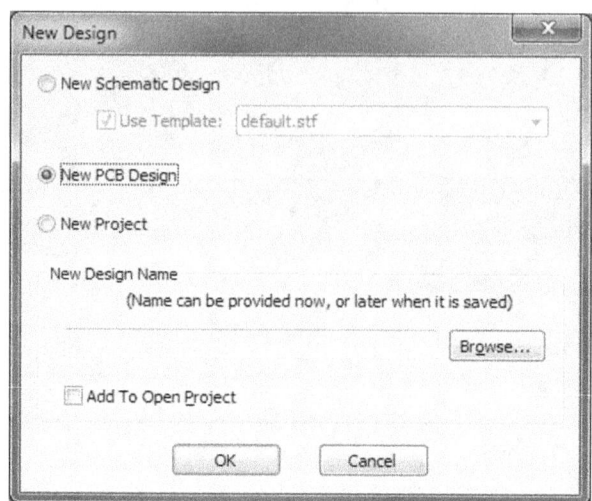

button.

The New PCB Board Wizard

First, you need to know that this will not result in adding components, pads, holes or traces. I know that I went to a lot of trouble to tell you that the schematic comes before the PCB but this process really sets up (or possibly resets) the project including, most importantly, the template file.

It is not absolutely necessary to run this wizard now, especially if you only want to use the schematic features, but it is better in the long run. The template files improves the probability of creating a project that is consistent in units, measurements, display and so forth. Things don't only look good, they work together better.

The New PCB Wizard begins with this screen. The vertical list on the left indicate your location in the board creation process. For example, in the illustration below you are at the Start location in the process. Note that in the lower left corner of the screen there is a Unit Price and Subtotal. This gives you an indication of how changes to the PCB affect the price. This is not so important now but becomes more significant when the board is close to being finished.

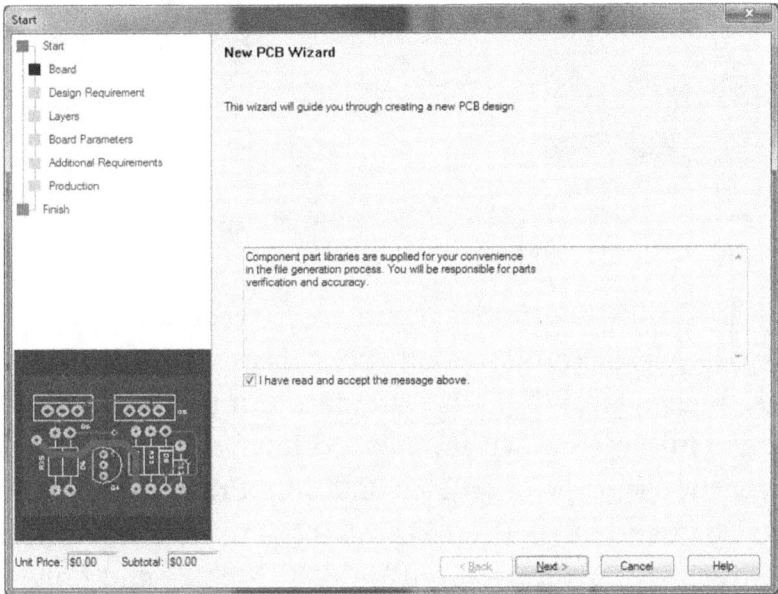

This window has changed since 1.5. The check box should read "I am an adult and can take responsibility for my own actions instead of complaining about this product which was freely provided to me."

It only takes a few to screw it up for everyone. America's current favorite passtime.

Select next.

Board Size Selection

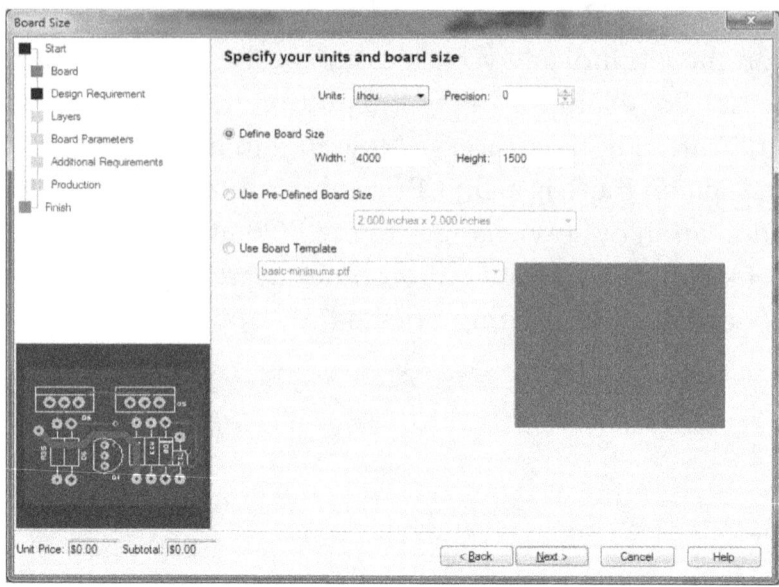

Specify Units determines if you are to work in Imperial (inches) or Metric (millimeters). For this design we will use inches. Inches are still the US standard but this will probably change to millimeters within the next few years.   So we select thou for thousandths of an inch (0.001".) Precision is set to 0 or $1/1,000^{th}$ of an inch. Printed circuit board manufacturing capabilities are currently about +/- .003" for size and hole location so selecting more accuracy would not be important. If a design has large pin count devices, say integrated circuits with, say, 50 pins per side and a pin pitch (spacing) of, say, . 0157" then you would want a Precision of 3 which is three significant digits after the decimal point.

This is so the error buildup from rounding is not significant over the 50 pins of one side.

For board size we will use 4 inches by 1.5 Inches. To do so we select the **Define Board Size** radio button. It is relatively easy to reset board size later so this is not critical unless you have a specific board size set in your requirements, as might be the case if you have a particular cabinet into which the product must fit.

First of all you may have a fixed board size per the requirements. If so ignore the next section.

A rough (very rough) selection is four components per square inch for through hole devices and ten components per square inch for surface mount boards. So, if you have 10 through hole devices and 20 surface mount devices a starting board might be:

10/4 = 2.5 sq. in for thru-hole devices

20/10 = 2 sq. in for surface mount devices

Keep in mind that one large component will make the above calculation pointless.

Usually the width and height is determined by connectors or other devices on a board.

When using surface mount devices it is tempting to double the devices per square inch if using both sides of the board. The real ratio is usually more like 20% more devices for a given board size. So, it would be 24 devices per square inch for surface mount devices if using both sides of the board for mounting components. The ratio can be improved but only with a great deal of extra effort.

I always select a larger board size so the components will fit inside the board during layout and reduce board size as the design matures.

Design Requirement Selection

The next screen determines the board specifications that will be used for this design.

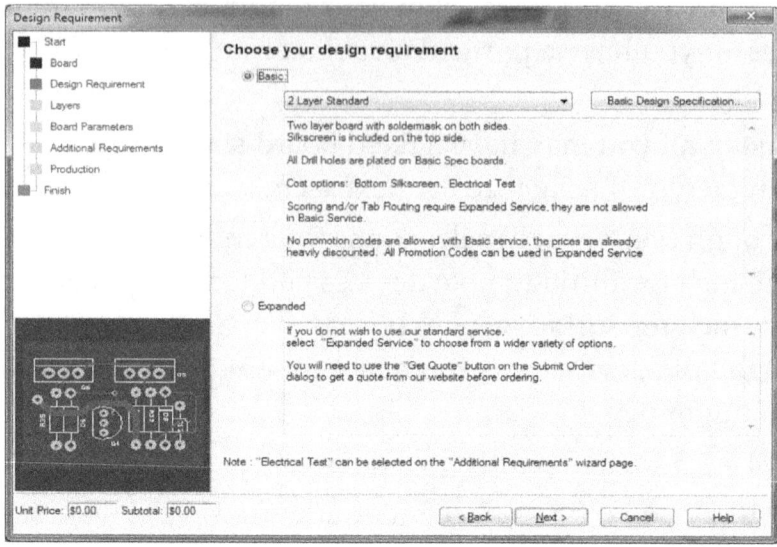

All holes plated through (non-plated holes available in custom service)

Non-plated slots and cutouts (plated slots and cutouts are available in custom service)

Green solder mask – Solder mask is the material placed on the board after the board is etched, drilled and the holes are plated. The silkscreen is added after the solder mask.

FR4 material – This is the current standard printed circuit board material. It is basically a piece of fiberglass with copper bonded to both sides. It will operate up to about 100 degrees C. The thickness of the copper is measured in ounces per square foot and stated as oz. For example 1oz. Copper is one ounce per square foot before etching[2].

[2]The process described is an etch process. Some circuit board houses use an additive process where the FR4 begins without copper and the copper plate is added.

White legend (silkscreen) – This is the writing that you see on printed circuit boards. Generally the line width for this writing should be .008" width or greater to produce good readability.

1 oz. Copper Inner layers. Up to 2 oz. Copper on outer layers – We are working with a two layer board, but the boards may be be multiple layers.

Trace Width and Spacing as small as 7/7 mils – This gives the minimum copper width of connections between devices.

Lead-Free HAL Solder finish – This is the visible finish of the copper layers. The solder mask allows some copper to be exposed to the air. This finish protects the exposed copper.

Hole Size Tolerance +/- 0.005" - For standard hole sizes such as .020" and .043" the finished hole size will be rather accurate, but for odd sized holes the finished hole might be .005" larger or smaller than required. If a hole has a minimum size it is best to increase the hole size by .005" just to be safe.

Overall thickness 0.062", 0.031" available on 2 layer – The standard board thickness is .062" or 1/16".

Does not include UL markings, 94V-O and date codes – These are regulatory and quality markings and may be added to the board for an increased price or larger quantities.

Layers Selection

This design will be a 2 layer board. All the default selections are acceptable as depicted below. The Automatic routing Bias selection determines layer direction when the autorouter is operating. Some auto routing problems can be solved by changing this from horizontal to vertical. It is primarily depending on large IC placement.

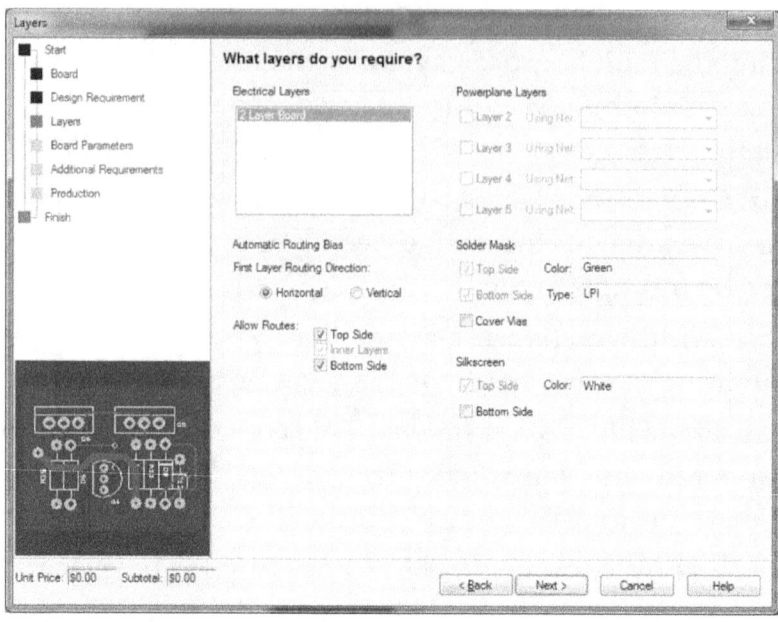

Board Parameters

The important thing to note is that these parameters are generally updated from the PCB design as it progresses. A very helpful feature is the Parameters Definitions button located in the upper right portion of the screen. It is extremely helpful for quick questions.

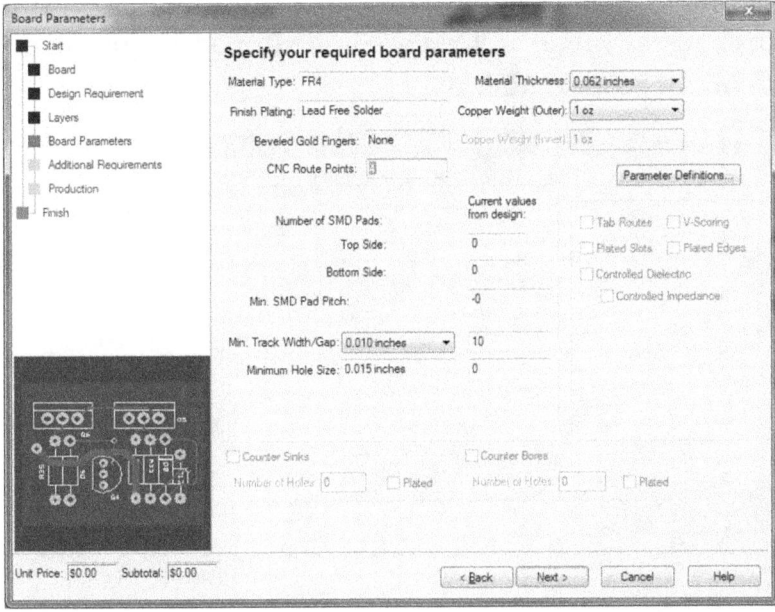

The required board parameters have been discussed above except Beveled Gold fingers.

Beveled Gold fingers – If the board is to be mounted in a card edge connector (this includes USB compatible boards) beveling (machining a ramp on each side of the PCB eases insertion force into the connector. If the connector pins are plated with gold they will be marginally more resistant to corrosion. I say marginally because gold is very soft and will scrape off after just a few insertions.

Additional Requirements

For this design it is best to leave this section open. I have submitted dozens of designs of various type to Advanced Circuits and have never had a board defect that would have been found by electrical testing. However, for high reliability designs such as Medical, Defense and similar programs this option might be essential.

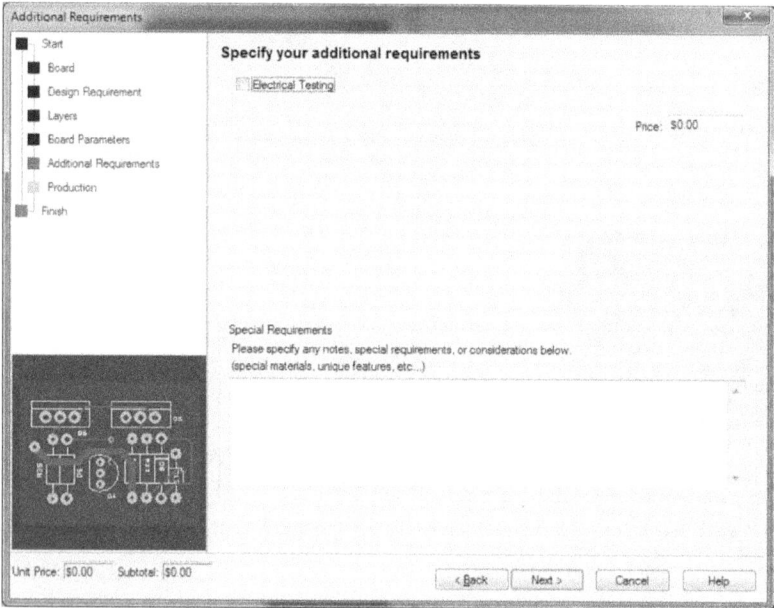

Board quantity and turnaround time

It may seem to early to get to this screen, but it is very useful to you. That price in the lower left corner of this screen gives you a good idea of how changes such as technology, number of layers, and quantity are affected by design decisions.

There are two very helpful and useful promotions available to Advanced Circuits customers. The first is 33each which provides prototype printed circuit boards at $33 each for a minimum quantity of four boards. The other is the barebones which is a one day service.

There are instructions on the web site for taking advantage of these prototyping options. The 33each promotion cuts the price indicated below by half.

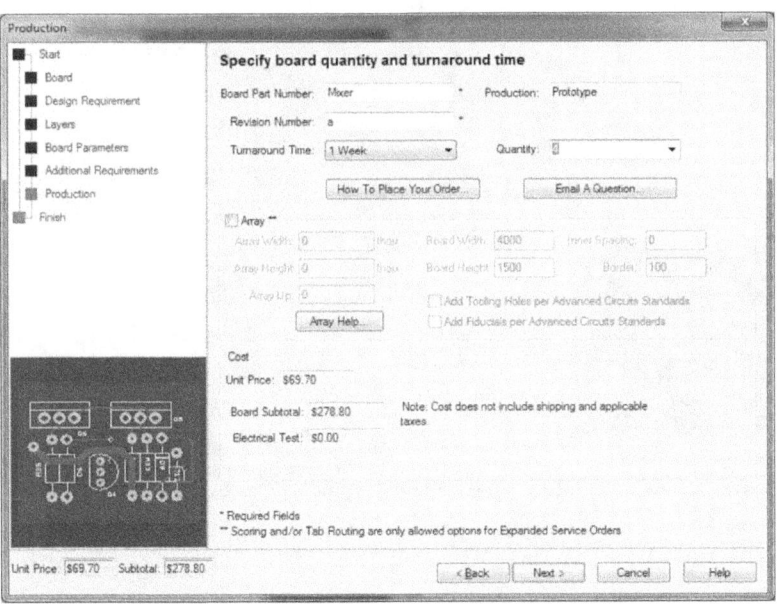

PCB Design Name

The project we are designing for this book is an audio mixer. It accepts a number of stereo source devices such as MP3 Players or Computers and combines the stereo sources to a single stereo output. So, for example, if you want to listen to your MP3 player and still hear your computer's sounds you can mix them and listen on your headphones (possibly at work) or sound system. This design would also be adequate if you want to create sound recordings. So I selected Mixer for the PCB Design name.

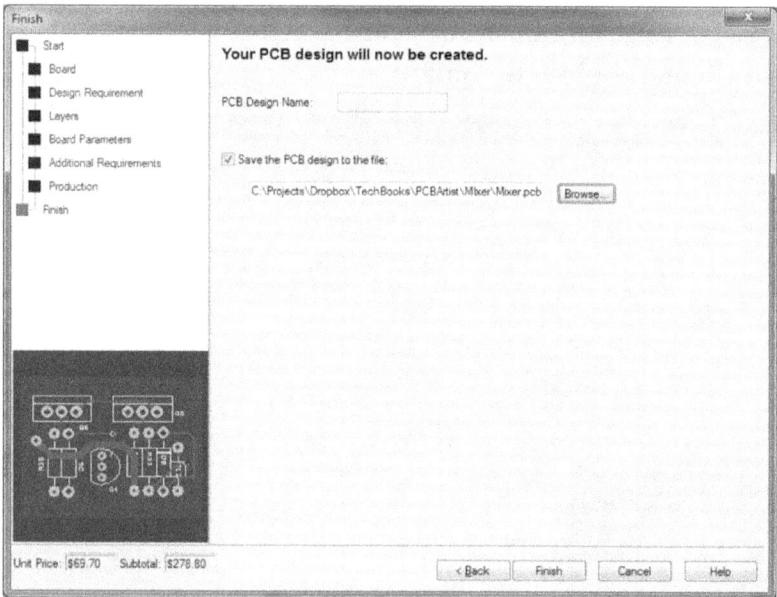

If you press finish you will be presented with a printed

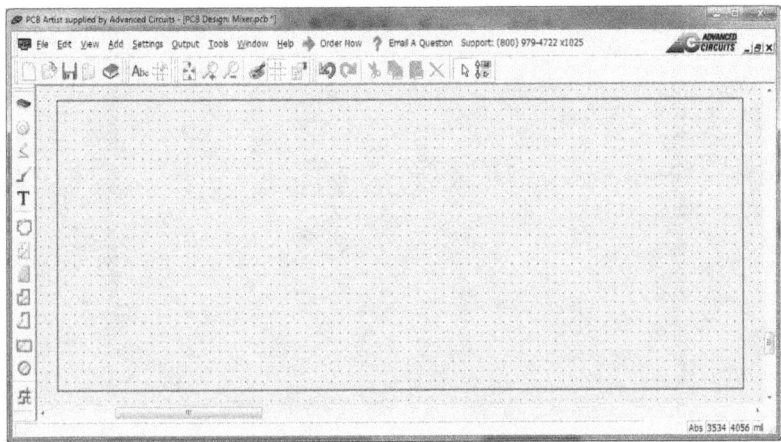

circuit board.

Notice the information in the lower right hand portion of the screen. Abs 8000 9000 and thou. The thou indicates that distances are measured in thousandths of an inch. The lower left corner of the PCB is located at location 8000 in the x plane (from the left side of the design plane) and 9000 in the y plane (from the bottom of the plane.) This is a good location as it allows components to be placed near the board when they are imported by the schematic package.

There is one important thing to check immediately. Go to Settings → Grid or press the Grids Icon

You will get the Grids Screen. I like to set the grid value to 50 thou when placing components and for the first autoroute. It gives spacing between components. If more precise placement is required the snap mode may be used to select Half Grid (0.025"), Quarter Grid(0.0125"), Tenth Grid (0.005") or Fourtieth Grid (0.00125"). Above all, never turn the grids off. If you do turn the grids off, turn them back on as soon as possible.

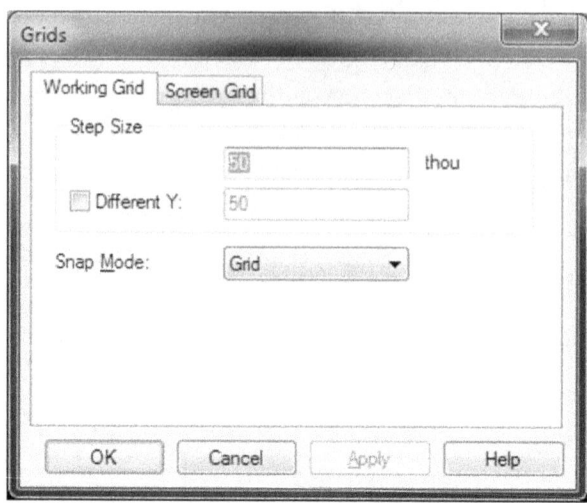

Having the grids on facilitates having the connection points connect up easily in either the library editor, schematic capture editor or printed circuit editor. This is especially true in the library editor because creating parts that are not on grid will be a source of endless aggravation.

Preferences

After the grids are set is a good time to set preferences. Selecting Settings → Preferences (or Cnt → F11) gives the Preferences screen. There are Tabs for classes of interaction. General → Security Copy Interval sets the time between background saves for a design. I use 10 minutes. The system will slow down, slightly, during a save but the first time power is lost or another application fails and data is lost you will appreciate a short interval.

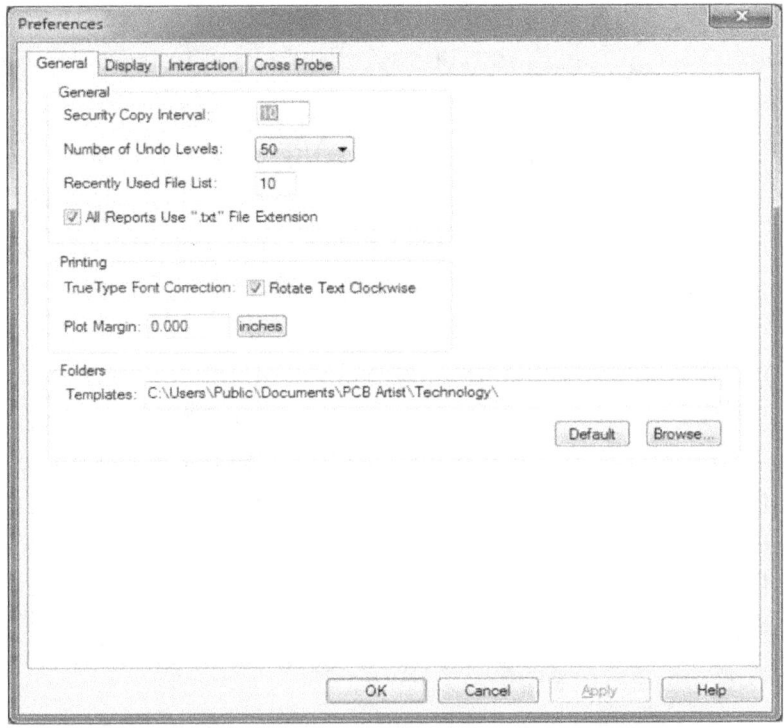

General → Number of Undo Levels sets the times you can perform a Cntl-x. I keep this high because I sometimes don't see the error of my ways for a long period of work, especially when manually routing a board.

General → Recently Used File List reminds you of the name of the name of the project before this one and the one before that.... I wish it would keep them all back to the beginning of time.

General → All Reports Use ".txt" file extension are the reports generated in the Output → Reports menu.

Printing → True Type Font Correction Rotate Text Clockwise is a good choice.

Printing → Plot Margin is usually handled by the printer and any value other than 0.000 would reduce your available printed space.

Folders → Templates contain starting points for various types of designs. They set items such as line width, route width, routing spacings and via size so that you can get these settings correct for designs that are similar. I always leave this at the default setting.

However, if you are shipping a copy of work to someone else you can save the settings for your project into a file of the same name and copy that template into the project directory. If you also add that file to the project files it will be connected associated with your project forever.

Display

This display refers to the computer's on screen display including mouse operation.

Drawing → Cursor gives the option of three cursor types. When working with the schematic or PCB editors the large cross option improves the process of lining up items, especially for neatness. I personally prefer either that or the Current Windows setting.

Drawing → Text Barring Character (when doubled) is an option I don't use because it is difficult to export at times with other packages. I currently prefer to use a trailing small case n to indicate a low true or negative true signal[3].

Drawing → View All on Opening Design forces a zoom all when opening a design. If you like to have a bit of a reminder as to where you were working last, this should not be checked.

---

[3]This is sometimes referred to as "Active Low" which is incorrect. Active low refers to origin of a current sink for a node. If a node is active low the current sink is an active device such as an open collector logic output or a discrete transistor collector.

Drawing → Detailed TrueType Text gives a trade off of beauty vs. accuracy. If this is checked TrueType fonts may not fit as shown on screen.

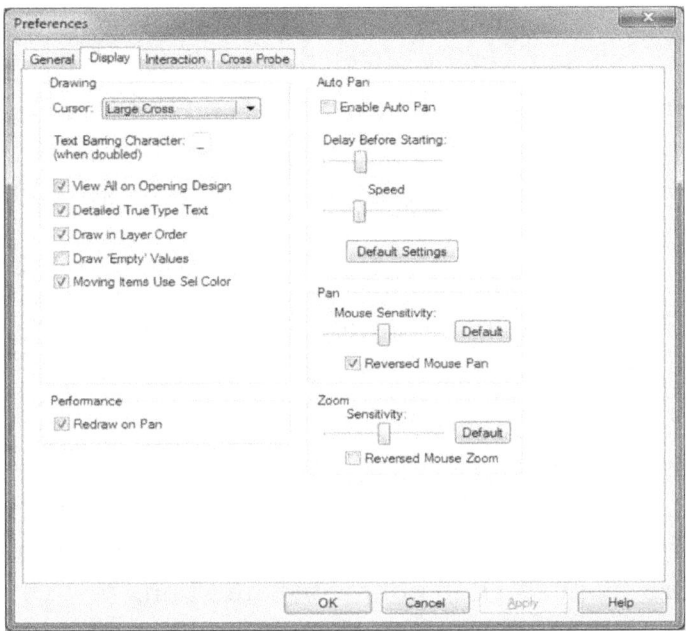

Drawing → Draw in Layer Order presents PCB drawings as if you are looking down at the PCB. The bottom most layer is drawn first and other layers are drawn on top of the previous layers. This works in conjunction with Settings → Colors → Settings and Highlight → Merge Colors which will allow a sort of x-ray vision where layers don't hide each other, but are more difficult to trace.

Drawing → Draw 'Empty' Values draw values that haven't yet been entered as question marks. This can become very messy and overrides otherwise intentional deletions.

Drawing → Moving Items Use Sel Color gives an excellent indication of what is moving. It is very common that the item(s) being moved are not the one(s) you intended. This option changes the color(s) while moving increasing the visibility of the items being affected.

Performance → Redraw on Pan depends on the user's computer, video card, RAM memory and patience. The display will always be most accurate when Redraw on Pan is checked.

Auto Pan → This series of settings are fully operator dependent. The auto pan feature allows the user to change the viewpoint of the display by moving the mouse cursor toward the left or right or top or bottom of the display.

It can take hours to get these settings to one's liking and they may have to be changed whenever system factors such as the mouse, the mouse driver, the video card and other system settings are changed. I do not enable auto pan.

Interaction

This screen standardizes the user interface to the software. Writing this book has been good for me and during the drudgery I found that I had the very first item in Peferences → Interaction set wrong for my purposes. I have been having trouble with library styles being overwritten when I import some symbols. I now select Add Component Keeps Library Style Sizes. At first I thought it was a good idea to use the project values and not check this box so that there is consistency in what is displayed and printed. However, now if there is a problem I go back and fix the library model.

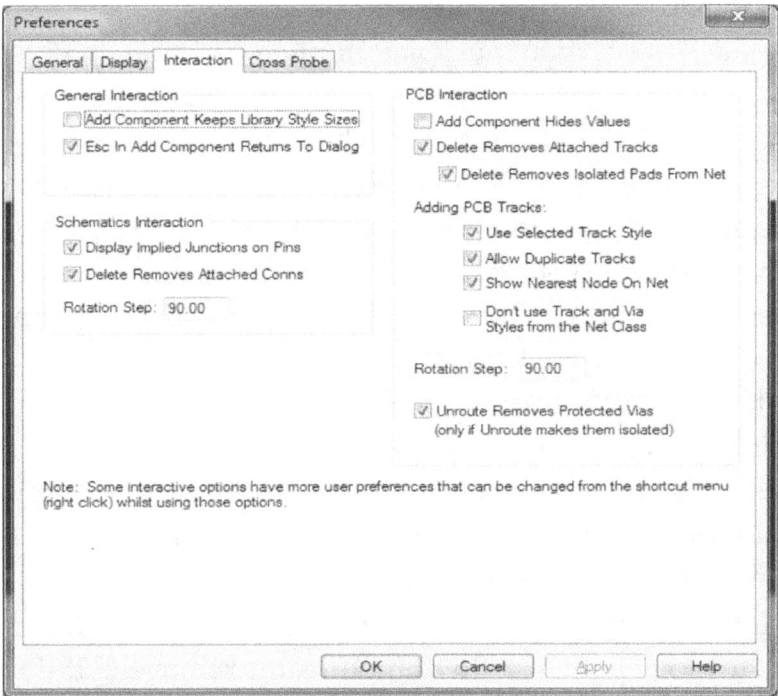

General Interaction

General Interaction → Add component Keeps Library Style Sizes affects the line width, color and other characteristics of components from the libraries.

I was recently bitten by this when I was working with it unchecked. I set a certain hole size for a component in the library and when I got the boards back the holes were the wrong size. Keep this checked and the library model will be used as it is drawn in the library, the way it was originally designed.

General Interaction → Esc in Add returns to dialog gives consistent operation with all the modules when checked. When escape is pressed the Add Component dialog remains instead of being removed entirely requiring the user to start adding components anew.

Schematics Interaction

Schematics Interaction-> Display Implied Junction on Pins assures that connections are displayed in a manner to make the connection obvious when checked.

Schematics Interaction-> Delete Removes Attached Conns will delete all connection to a component if the component is removed. If three components are connected on a net the connections to the deleted component are removed and the other connections are left intact.

PCB Interaction

PCB Interaction → Add Component Hides Values makes all values invisible when a component is added to the design. The author sets value visibility in the library. Value visibility can be altered after the component is added.

PCB Interaction → Delete Removes Attached Tracks deletes all tracks to a component if the component is removed. If three components are connected on a net the tracks to the deleted component are removed and the other tracks are left intact.

PCB Interaction → Delete Removes Isolated Pads From Net assures that remote pads (possibly on the other side of the board) are also removed from the Net.

Adding PCB Tracks

Adding PCB Tracks → Use Selected Track Style is particularly useful when routing boards that have different styles of tracks for different nets. This includes most boards as power and ground nets are often larger than signal nets.

Adding PCB Tracks → Allow Duplicate Tracks allows you to create an alternative routing without deleting the existing route. This is useful when manually routing the PCB and alternative routes need to be tried without loosing the original routes.

Adding PCB Tracks→Show Nearest Node On Net provides guidance while routing, especially when a long route requires that the display be panned during routing. In this case the net line jumps from one available node to the next as the cursor is moved along. This becomes more handy as the board becomes larger, but there is a performance penalty. I always use it but I have a four processor Pentium and 8 Gigabytes of memory. On my laptop I turn this off.

Adding PCB Tracks → Don't use Track and Via Styles from the Net Class allows the settings in Settings → styles to be overridden. This should be used with care primarily because changes in the net style settings may be made globally, but setting this option requires reworking all nets independently.

Rotation Step → Normally this is left at 90 degrees, but when a component will only route when placed at an odd angle. This does not affect the angle(s) available during track routing.

Unroute Removes Protected Vias → Protected vias are protected from being deleted when changing layers by means of the Cntl-L changing of layers. If a via or vias are not protected changing layers such as with a Cntl-L will delete unneeded vias.

Cross Probe

Cross probe allows selection of a device in the PCB editor to be highlighted in the Schematic editor and vice versa. This is particularly useful during the component placement phase of PCB layout.

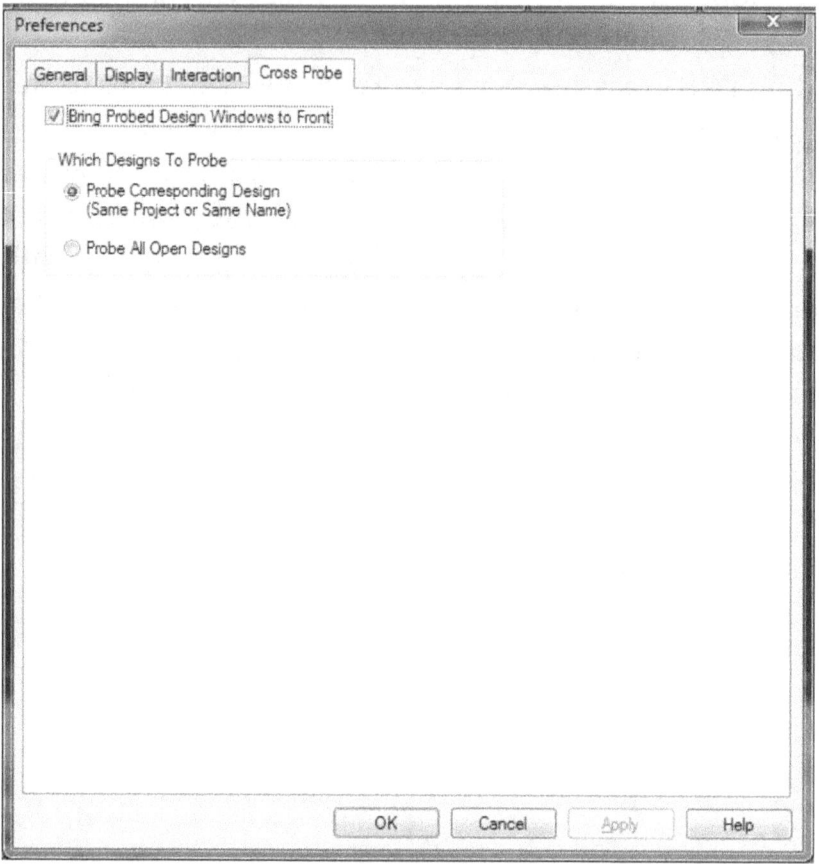

Schematic Capture

A schematic consists of a diagram that indicates connections between components. The old methods of creating line art were time consuming, costly, prone to errors and easily damaged or destroyed.

Schematic Capture software, like PCB Artist allows this information to be entered into a computer. Once in digital form the schematics may easily be transmitted or reproduced. And with the proper procedures should remain as accurate as the original product for an indefinite period.

Connections

Schematic capture is the process of creating a diagram that represents the various connections of an electrical or electronic circuit. For this document the United States standard, as opposed to the European standard, component bodies and electronic layout standards will be used. The connections and functionality are equivalent for each drawing style.

The resistors above have two attachment points. Electrons are always in motion, but that motion is Brownian (random) in nature. Brownian motion is the random motion of particles, including electrons, without any total displacement. Browian motion is truly elastic in that energy doesn't change form.

Current is the orderly flow of electrons from one point in a circuit to another point. In any volume of matter the net movement of electrons is zero. When a current is present the movement of electrons follows the route of least resistance (a circuit) from a power source and back that that same power source.

Current flow can transport energy from one location to another with minimal losses as in the transmission of power throughout the public power system. Current flow can be changed from electrical energy to heat, or light or radio wave or motion.

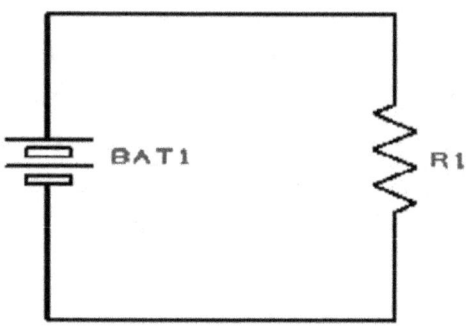

The drawing above illustrates a very simple circuit. The battery provides power by means of a chemical reaction. The current leaves the battery, say from the top, and travels through the conductor (wire) to the top of the resistor. The current then goes through the resistor. After leaving the resistor the current then reenters the battery. In simple terms current goes from a power source (the battery) through a load (the resistor) and returns to the power source.

There are two schools of though regarding current flow. Engineers generally follow the Franklin model of current flow being from the positive terminal of a power source to the negative terminal of that same power source. This is often referred to a 'current flow.' Technicians generally are taught that current flows from the negative terminal of a power source to the positive terminal of a power source or electron flow. The result is the same, only the terminology is different.

One whom creates schematics and printed circuit board for others should be ambipolaric, that is, be conversant in current flow or electron flow.

The figures above depict a through hole resistor and a surface mount resistor. The through hole resistor has metallic leads extending from the body. These metallic leads are the conductors which are connected to the circuit. The surface mount resistor has leads that are a part of the body. In either case the component is attached to the Printed Circuit Board (PCB) by soldering the leads to conductors on the PCB.

Nearly all components in a schematic will have two or more attachment points so that electrical current can move through the component. There are exceptions such as connectors and test points. In each case external connections create the connections.

In this drawing below another load has been added. R2 is connected across the battery like R1. There is a line from the top of R2 to the line that connects the top of R1 and the top of the battery. There is a dot at that intersection. That dot indicates there is a connection at that point. The top line of R2 does not have a dot indicating that there is no connection.

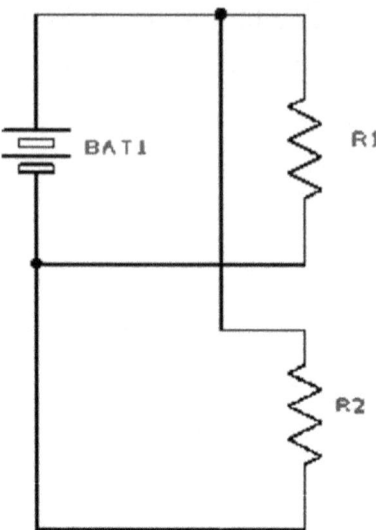

The connection to the top of R2 goes across the connection between the bottom of R1 and the bottom of the battery. There is no dot at the junction of the two lines and thus there is no connection.

Components

Components are the work horses of a circuit. In the drawing above the components are a battery and two resistors. The battery provides power and the resistors consume the power in the form of heat. If a motor were used instead of a resistor rotational motion could be produced. If a lamp or LED were used instead of a resistor we could produce light.

Some components are quite complex (microprocessors) and others are quite numerous (resistors.) Since we now have the information in a computer we can create collections of similar components and provide these collections in the form of libraries. A number of libraries are provided with PCB Artist. However, it is likely that you will need to create additional components.

There is a library editor for that purpose. Note that there were two forms of resistor depicted above. One was the actual physical outline of a component (the European standard) and the other was a squiggly line representation (the US Standard) of that physical component.

Editors are provided for both schematic symbols and PCB symbols. These editors are ingenuously inter-connected to provide a very consistent interface to you the user.

The PCB Artist Libraries

The PCB Artist library editor consists of a Schematic Symbol editor, a PCB Symbol editor and a Component editor. PCB Artist combines the two editors creating Components that contain the Schematic and PCB symbols. This results in less build time (one editor,) a constant user interface and predictable results.

With the advent of PCB Artist 2.0, libraries may or may not include SPICE information. Spice information may be added to the schematic at any time, but adding the information to the library means that the information may be added once for every use of the product.

From a quality point of view having a single repository for information is preferable to multiple sources of information. Synchronization of multiple sources of information is inherently difficult and considered to be a prime source of quality control issues, including inventory management, component qualification and design integrity.

This book will demonstrate a technique of using the values in the component models to create an accurate and useful bill of materials directly from the schematic and PCB from which parts can be ordered. This is usually a separate step for other schematic and printed circuit editors. It should be noted that certain components will not be included in the bill of materials generated from PCB Artist such as mounting screws and standoffs. These are actually not a part of the printed circuit board and should be included in the mechanical drawings.

To create a component one must have the schematic symbol and PCB symbol available in their respective editors. We are going to create these symbols for an OPA2322 operational amplifier. The first step is to get a copy of the most recent data sheet. This part is available from Texas Instruments. If you go to the TI.com web site and search for OPA2322AIDR you will get a link to the component web page. You may get a copy by means of an HTML link under the Datasheet link. Download the datasheet and open it.

As noted elsewhere in this book, I often have better success doing a Google search with the part number and type of request. I typed **Opa2322 datasheet** into my browser and received a direct link. Manufacturer's web pages often require extended time to wade through their marketing information to get to the product they are selling. Weird, but true.

This type semiconductor is called an opamp which is short for OPerational AMPlifier. Note that the OPA322 is a single opamp, the OPA2322 is two opamps in one package and the OPA4322 is four opamps in one package. Even though we will need four opamps the use of the dual package will maximize circuit density (reduce board size) without increasing layout difficulty.

This drawing is an excerpt from the TI data sheet. And shows the connections for signals, power and ground for the IC. Each opamp has two inputs and one output all of which are separate. There is also a V+ and V- pin. These pins are shared between the opamps. This sharing can be problematic in schematic and PCB layout as it represents two connections (one to each opamp) with only one pin.

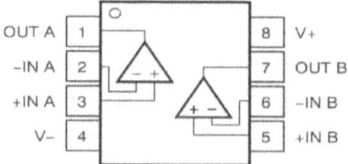

One method for solving this conundrum is to draw both parts in a single package. And this is the approach we will take here. This approach has the advantage that it will assure that the components are selected (placed) in the schematic section of layout instead of the layout section. This assures that components which should be placed near each other (such as the two sections of an opamp when there are a number of opamps of the same type in a circuit) are assigned together with automated layout tools.

The steps to create this schematic symbol are covered in the section on schematic symbols. The design philosophy allows a remarkably small number of schematic symbols compared to other PCB layout design software packages, especially for complicated schematic design symbols.

The next step is to create a PCB symbol (also commonly known as a footprint) for the part. This part was selected because it is a surface mount device. There are SO-8 PCB Symbols in the libraries. You can find prepared symbols by using the find option. This author most often uses components from the *prolib* library as they are usually the most accurate. We will create a unique SO-8 library symbol.

The steps to create this PCB symbol are covered in the section on PCB symbols.

Creating Surface Mount Schematic Symbols

The most important step to creating a successful schematic symbol is getting a current and accurate data sheet. Literally millions of electronic components have been created over the years and many have similar names and function. It is essential that you start with the correct data sheet from the correct manufacturer. It is also important that it be the most recent design data sheet.

It is also a good idea to save a copy of the datasheet used to create a product so that you may have a good idea of the baseline product when there are changes.

The best place to get these data sheets is from the manufacturer. This is sometimes difficult for a variety of reasons so an alternative method is to procure the datasheets from the retailers (they want to be called wholesalers, but look at the prices) that have copies or links to the most recent data sheet. Mouser (www.mouser.com) , Jameco (www.jameco.com) and Digikey (www.digikey.com) will handle the majority of items and have links to current datasheets.

Still, when possible, it is best to get the datasheet directly from the manufacturer. I've found that it is often easier to find a manufacturer's datasheet through Google than through their website. The trick here is to add the manufactuer's URL to the search term.

We will begin by creating a schematic drawing for the OPA2322 component. If you go to the Mouser.com website and type in OPA2322 in the search window you will get a

number of results. The same part is provided in many
different packages. Therefore it is essential that you get the
complete part number. In our case the full part number is
OPA2322AIDR.

The D in this part number
indicates the D or SO8 or SO-8
package. The package layout is
represented to the right. We will
create this schematic model as
one schematic model as
discussed in the Components
section.

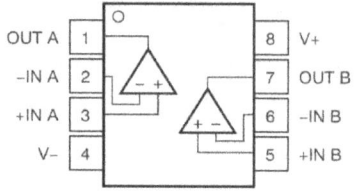

D, DGK, AND DCU PACKAGES: OPA2170
MSOP-8, SO-8, AND VSSOP-8
(TOP VIEW)

The Library Manager

Schematic symbols, PCB symbols and components are
edited with the Library manager. The library manager is
started by pressing CNT-L or Alt-F-> L or choosing Library
from the File pull down menu (File → Library).

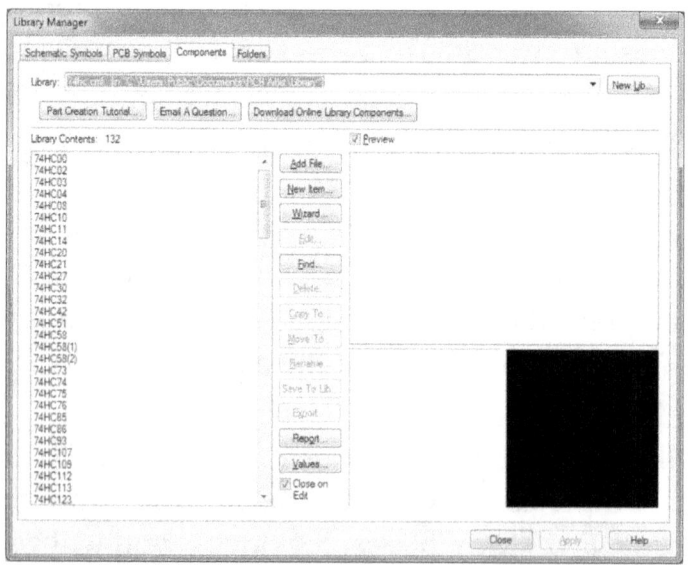

You have the option of putting your models into the existing libraries, but I don't recommend this practice. Whenever the software is updated the libraries may be overwritten. In this case you will loose the symbols you have created. For that reason it is suggested that you create your own libraries. You may put these libraries in the same directory as the PCB Artist libraries. When the software is updated the new libraries will not overwrite your libraries if you use a unique name.

Creating A New Library

To create a new unique library you create click on the folders tab in Library Manager.

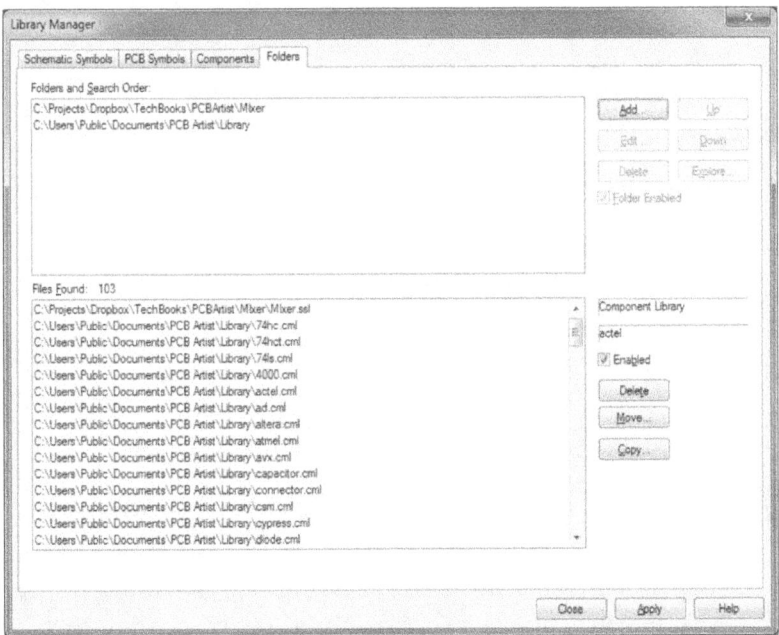

You than click on add which will give you a Browse box. You can then either type in the folder or click browse. In this case I'm using the same folder as the mixer project. Generally you would want to choose a single location to which all your projects will point. I often use c:\projects\library\ for my common libraries.

Schematic Symbols

Then you select the Schematic Symbols tab in the File → libraries window. We will create a new library by clicking the New Lib... box.

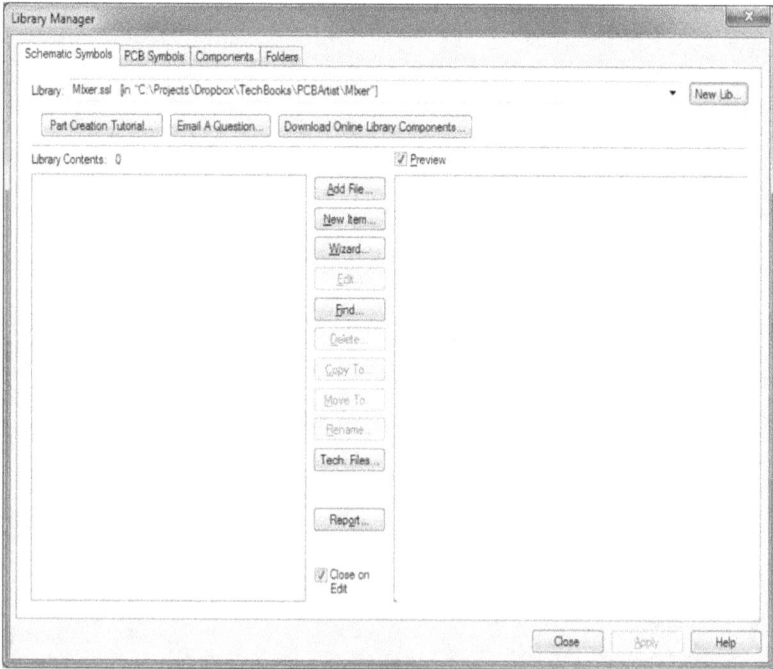

I elected to put the new file in the Mixer directory so that I may include the library with the schematic and printed circuit created here.

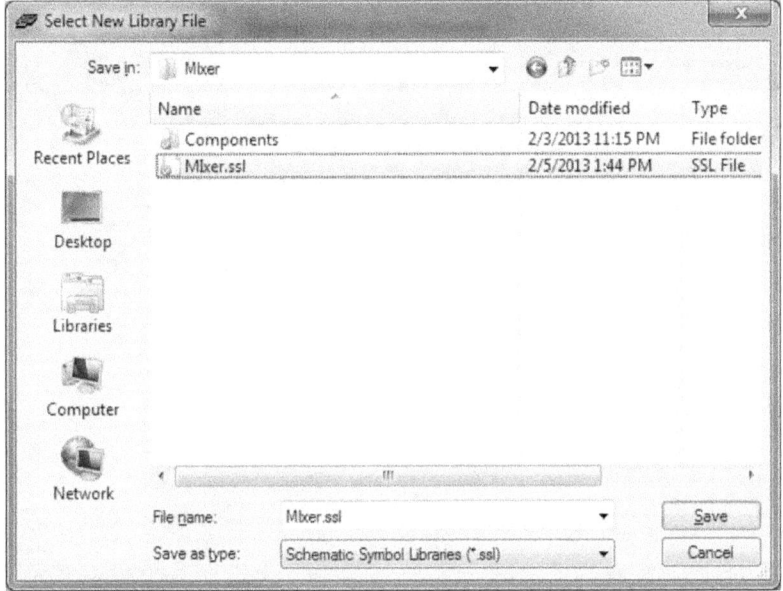

Let's say you don't want to create a new schematic symbol. You can check if there is an existing schematic symbol by pressing search which gives a find window.

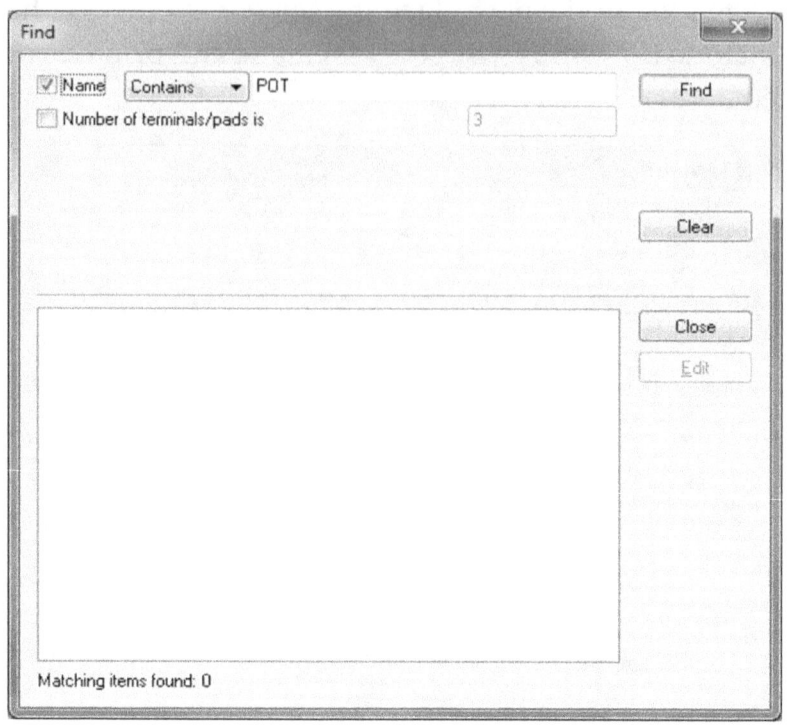

I entered opamp and selected Contains from the pull down menu. The possibilities are Contains, Starts With and Is Exactly. I use Contains since it is difficult to determine how the name might be constructed by others. If you know the number of connections (this is usually the number of pins, but not necessarily. There are often more pins than connections for an IC.

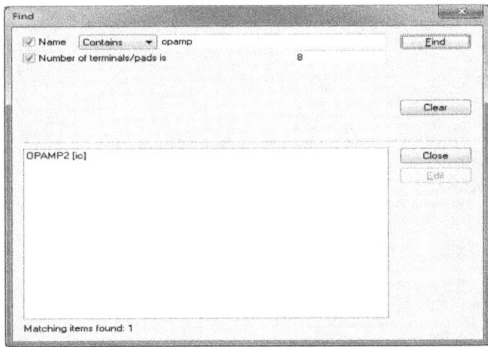

Note that I got only one response from the search. I then selected the component and got preview that component. (Not that the preview box must be checked to review the schematic symbol in advance.

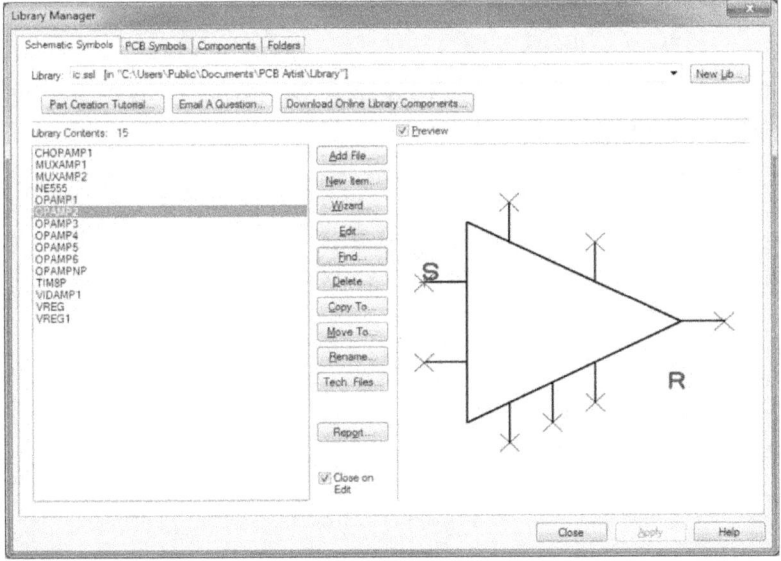

This is not close to the way we want to draw this part so we press the library pull down tab and select our newly formed Mixer library.

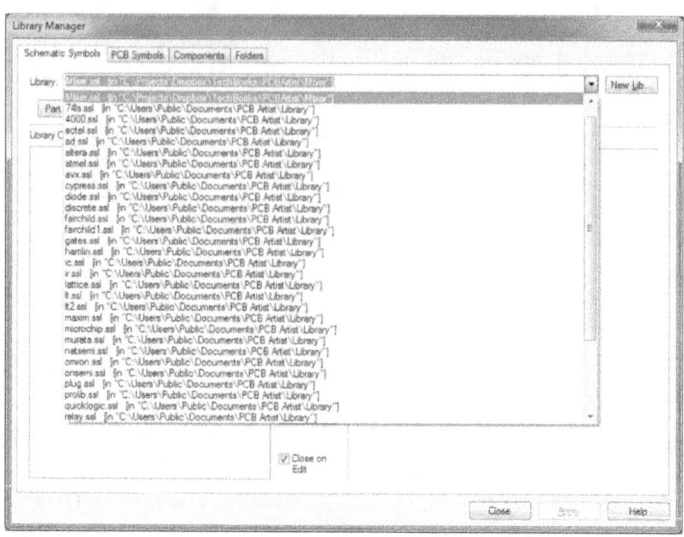

Schematic Symbol Wizard

Now we are ready to create a new Schematic symbol we will select the Wizard.

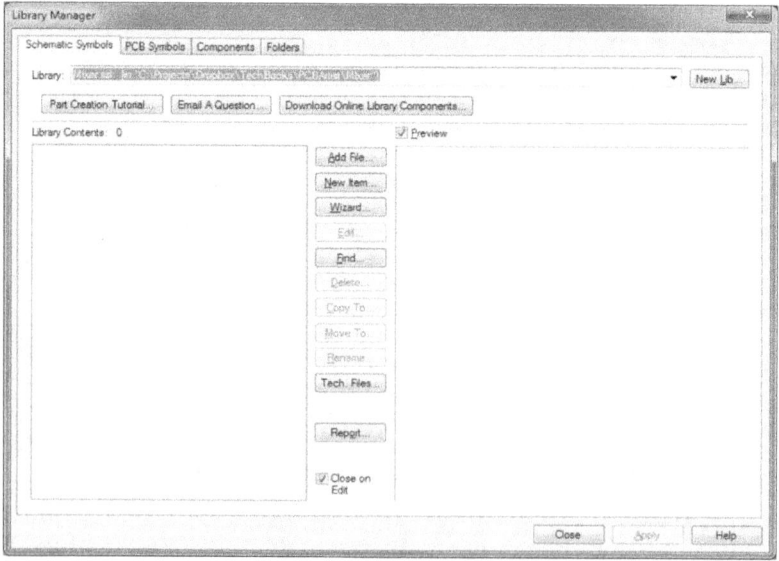

After selecting the wizard we will start the Symbol Wizard process with the following screen.

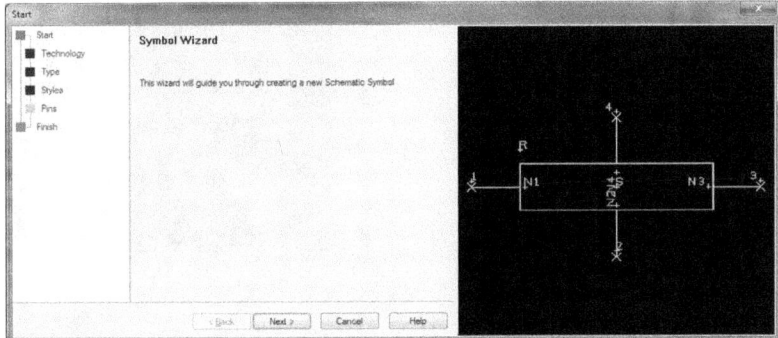

Press next.

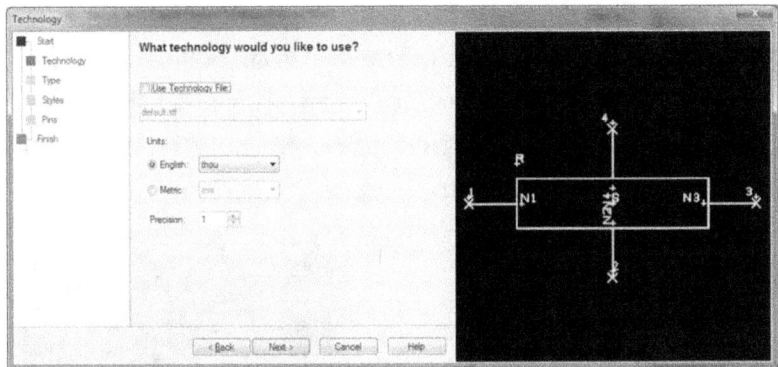

You can fill in the box like this. Don't worry the IC you see is not a representation of what will be created. This is the Technology box. Thou means that units will be one thousandths of an inch (0.001"). So 1 thou is equivalent to 0.001".

The next box is the Symbol Type box. We will be drawing a couple of triangles but the triangles seen here are not what we need. We will draw our own triangles. The lower Rectangle box has pins out the sides and the top and the bottom of the symbol. This is the symbol we need.

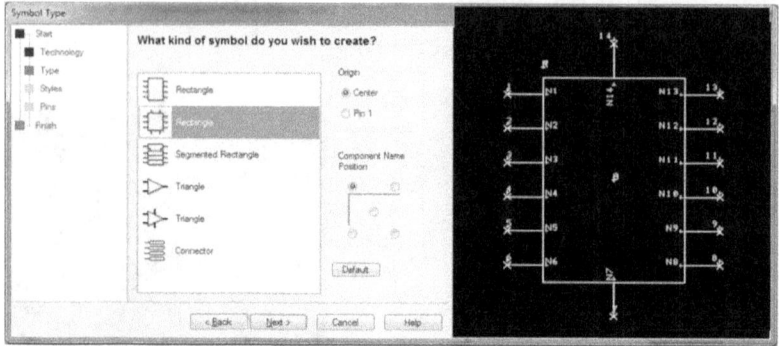

Press next and you get the Styles menu. The only common change to this menu is the Line Width: which should be changed from 5 to 8. This makes the schematic easier to read when printed on a standard printer – as opposed to being plotted on a plotter. Take note of the x shapes at the end of the pins. This is the point to which lines will be attached. All the rest is added to make the function of the schematic clear.

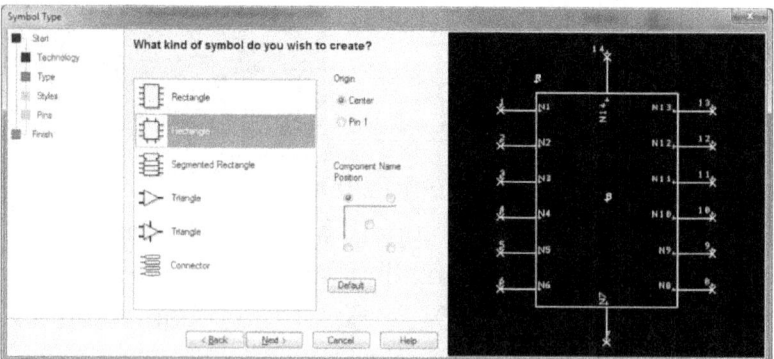

Press next and we get the pins window. We will have a total of 8 pins on our device. There will be four pins on the left, two pins on the right and one each on the top and bottom.

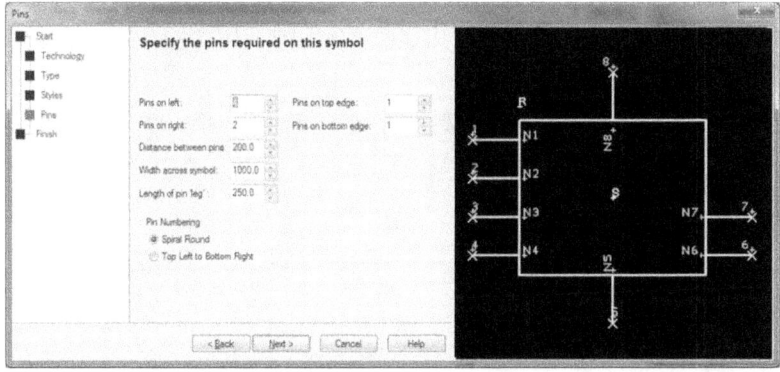

Distance between pins:	200
Width across symbol:	1000
Length of "leg":	200

For pin numbering you can press the radio buttons to determine what will happen.

Press next and you will see the Finish window. Insert the Symbol Name OPA2322 and check the boxes for Save the symbol to the library and Edit the symbol now.

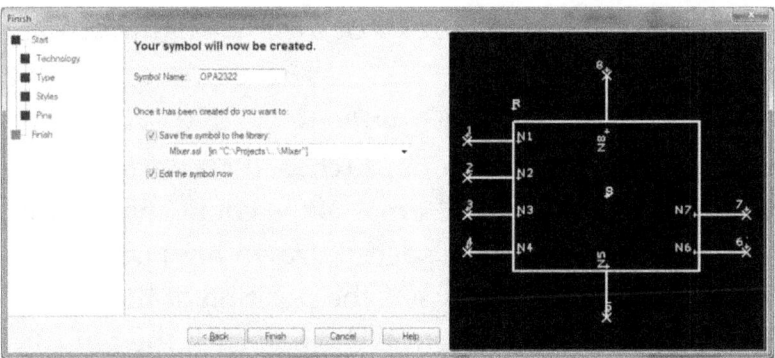

Press next and you are placed in the editor.

We will make a number of changes to this symbol in the editor after we exit the wizard.

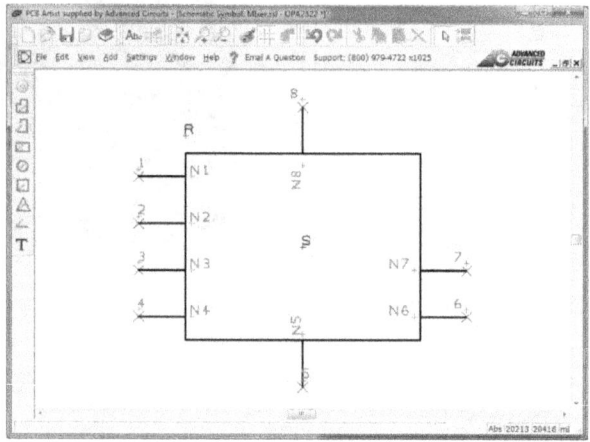

The drawing from the TI datasheet is illuminating but there is a better way to represent the part for end user understanding. It is advantageous to have the inputs on the left and the outputs to the right. Schematics are often drawn with the positive power entering the symbol from the top and the negative power entering the symbol from the bottom.

First press Alt-V → A to center the component in the viewing area and you should see a representation as below. Remember that the x's at the end of the line are the actual connection point.

Left and right pin placement

If we view a close up of a pin 4 (chosen randomly) there are two crosses and one X. The X marks the connection point. To make a connection a wire (a special drawn line) must end on the X. The cross above the X sets the position of the pin number. The cross to the right is the name for the pin.

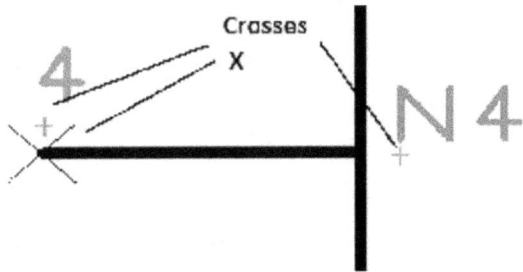

If we select the cross above the X and execute a right click we get a action dialog box. If we select properties in the Properties box we get a Properties dialog box like the one below. Be sure the Text tab is selected.

The Alignment: pull down box has three choices Left (shown here) Right and Center. If the pin is on the left side of a symbol Left alignment should be selected. If the pin is on the right side of the symbol

Right Alignment should be selected. If the pin is on the top or bottom of the symbol Center Alignment should be selected.

Name Alignment: can be selected the same way.

The first thing I'll change is right click on the N5 pin and select either angle 0 or press R until the Name is reading vertically. This part will be understood in the circuit better if the opamps are arranged over each other so the pins are moved to this configuration. The pins are now 2, 3, 5, 6 down the left side; 1 and 7 down the right side; 8 on top and 4 on bottom. This renumbering could be handled at the Component stage but I'm demonstrating pin manipulation.

To straighten this out we first place the X marks on the pin ends. This is accomplished by selecting the cross. Then we can select the Names by clicking on them while being careful not to select more than the name. The name alignment and location are set. The pin numbers are then placed in the same manner. When finished the model should look like this. Also

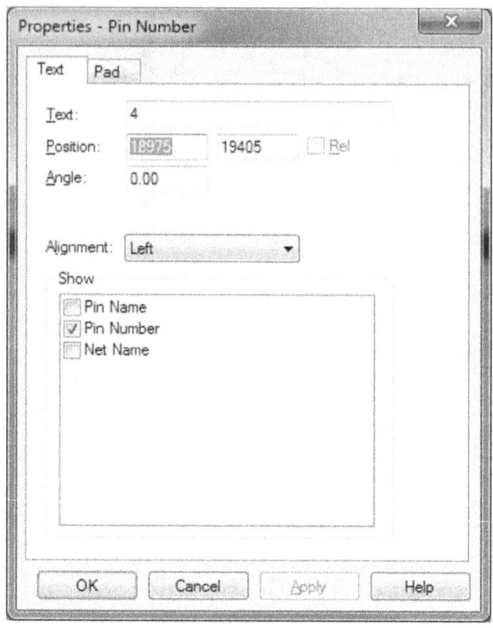

note that the R has been moved. The R is the place keeper for the reference number. Originally it was up in the air and to the left. It has been moved to the upper right of the schematic symbol. This is where the reference designator will be placed when first placed on a schematic.

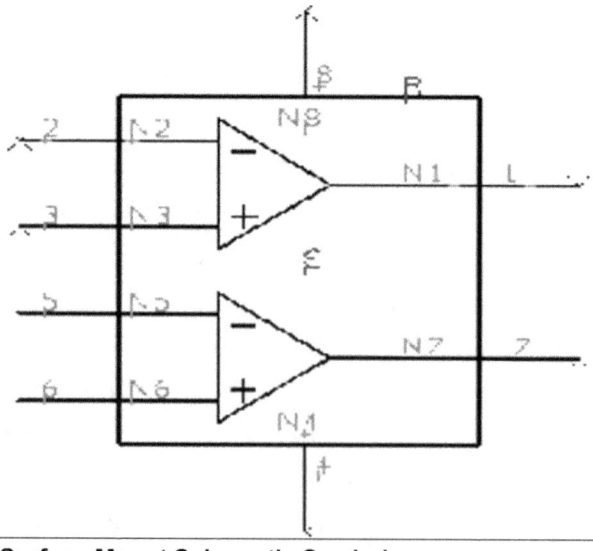

Alt-F → S or Cnt-S will save the part. Note that you have the option of putting it in any Library and using any unique name for the symbol.

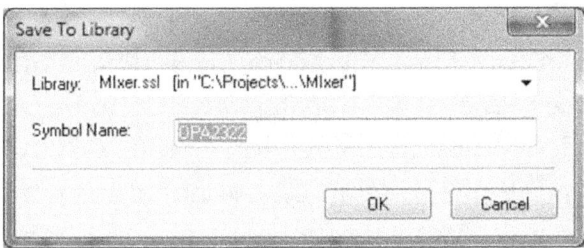

Creating Through Hole Schematic Symbols

This section could be called complicated Schematic Symbol Creation. The name only indicates that it is to be used with the Through Hole PCB Symbol Creation chapter. A through hole component symbol is sometimes, but not always, a bit more difficult to create. For example, there are times when the component is not mounted to the board as would be the case with, say, a washing machine switch or a stereo potentiometer.

The sample component is a dual potentiometer usually referred to as a pot. The pot selected is used as stereo volume controls in this circuit.

A datasheet for this component can be obtained at

http://www.bitechnologies.com/pdfs/p140.pdf

The specific model number is P140KV1-F25BR10K. This component has nine legs that protrude through the board but only six of those legs are connected to the circuit. This is a case where there are more PCB holes than schematic connections. The other three holes are for physical mounting purposes.

A note regarding Advanced Circuits prototypes ordering is that all holes must be plated for prototype processing. This assumption allows Advanced Circuits to provide their prototypes at reasonable prices.

This offers a distinct advantage to the layout engineer. I take advantage of this and make the pads on both sides of the board the same diameter as the screw head assuring that I will not create a disadvantageous circuit situation (short) when the product is assembled.

We start by typing Cnt-L to open the Library manager and click on the Schematic Symbols tab.

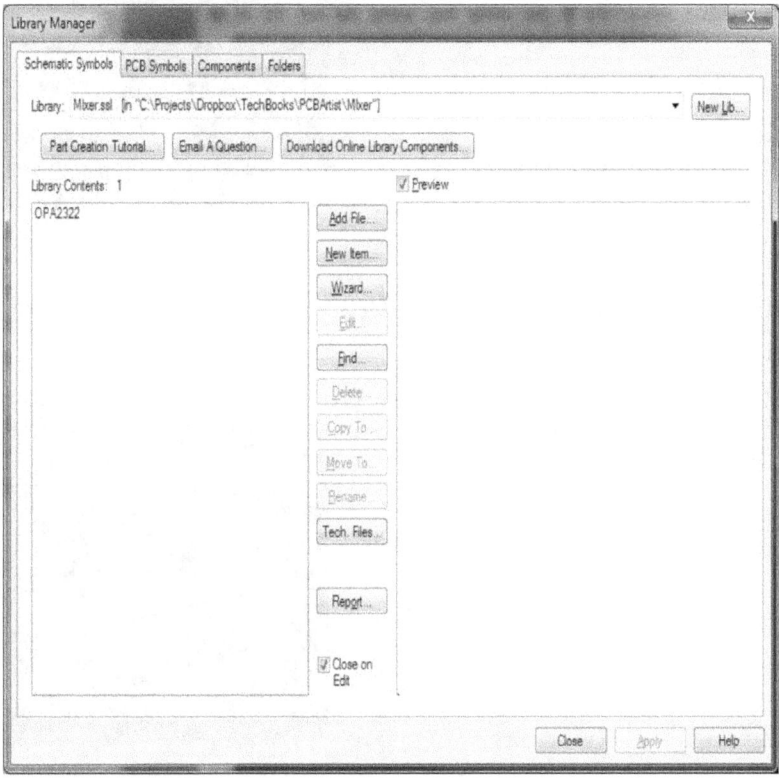

We select the Wizard to begin development of the component.

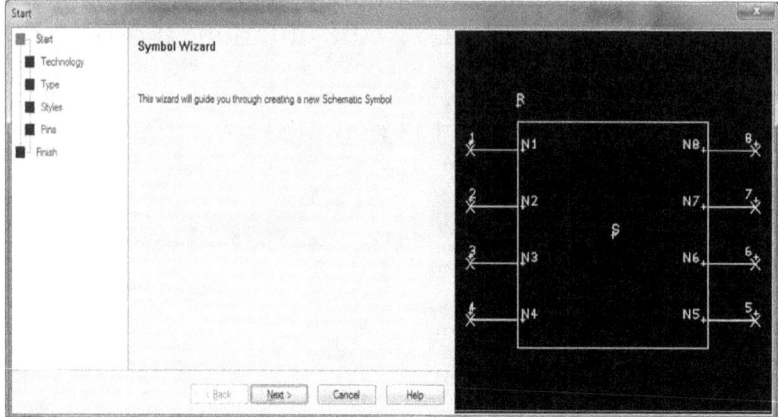

We select Next to begin the technology section where we select English: → thou.

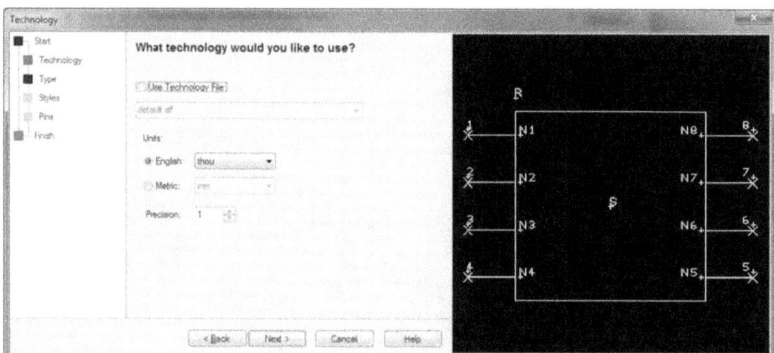

We select Next to begin the Type section. This time we select Rectangle as there will be four connections on the left side and two connections on the right side of the symbol.

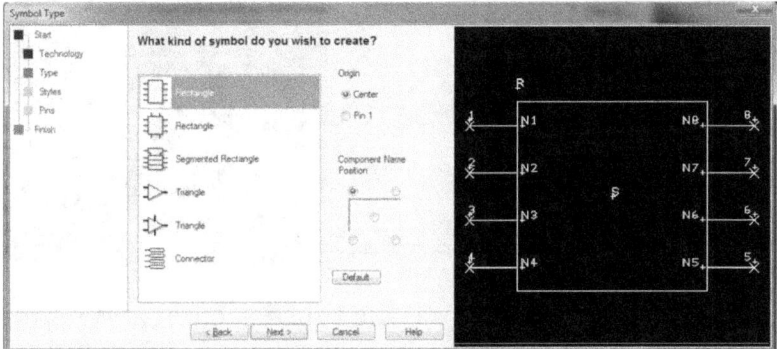

Then we select Next to begin the Styles section. We only change the Width: to 8 in the styles section.

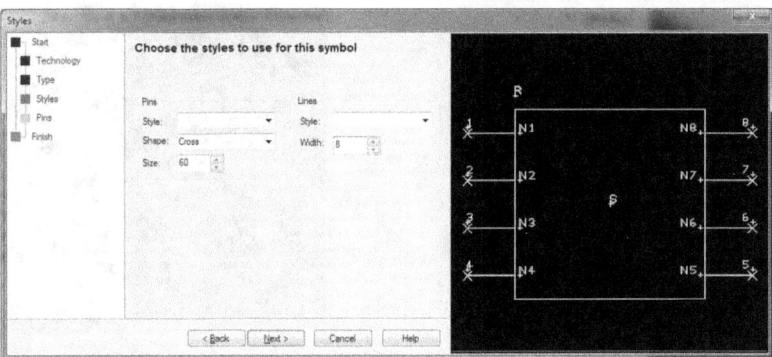

We select next in the Pins section and set the wizard for four pins on the left and two pins on the right.

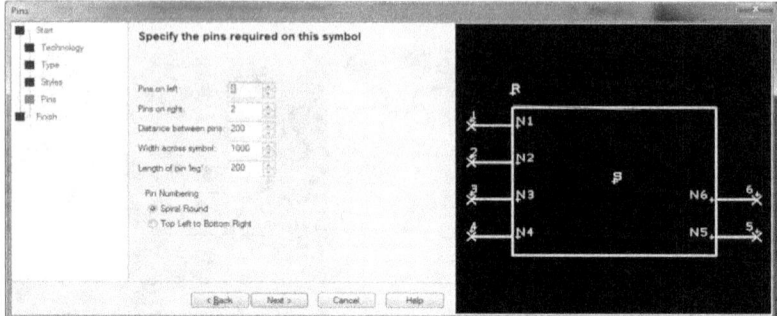

We then select Next to go to the finish section and name the part DualPot.

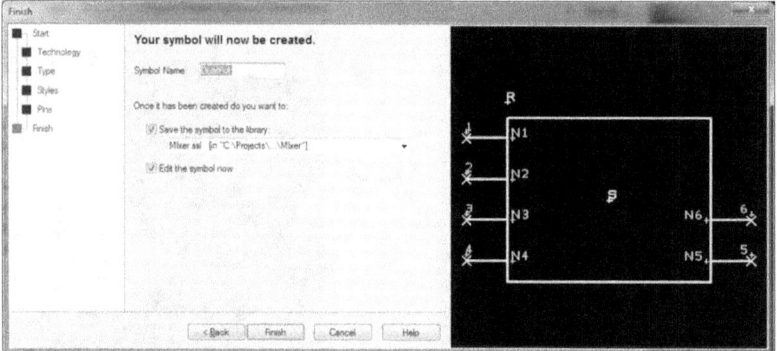

Select finish and we get a box type model. This is really a starting point. Press Alt-V A to center the part on the screen and we will begin modifying the part for our purposes.

Select the box and press delete removing the box.

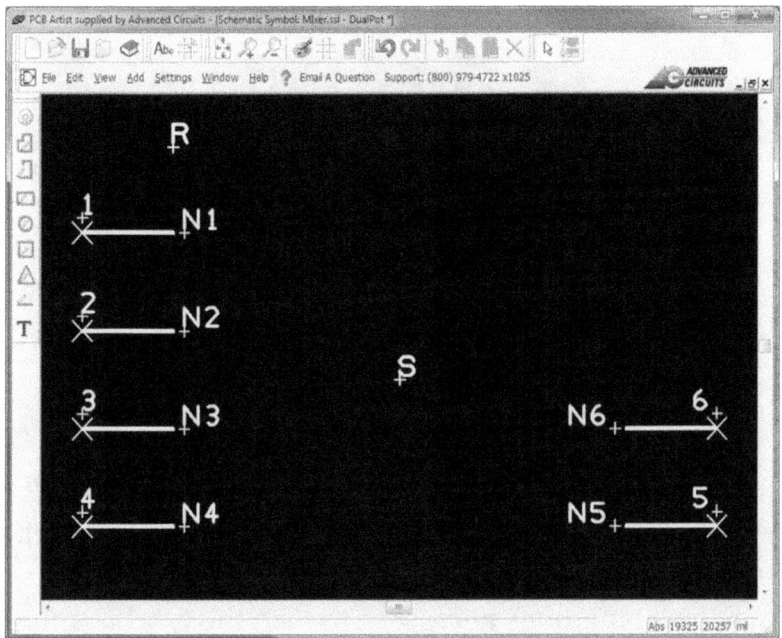

Select the Grids Icon   and select the Working Grid Tab. Then set the Grids to 25 thou and the snap mode to Grid.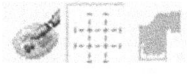

Select the Screen Grid tab and set the Screen Grid to Same as working Grid.

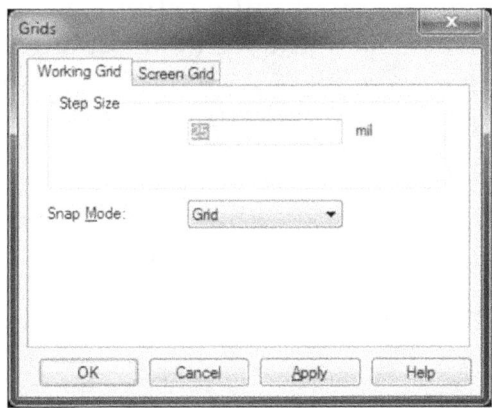

Assure that the grid display is enabled.

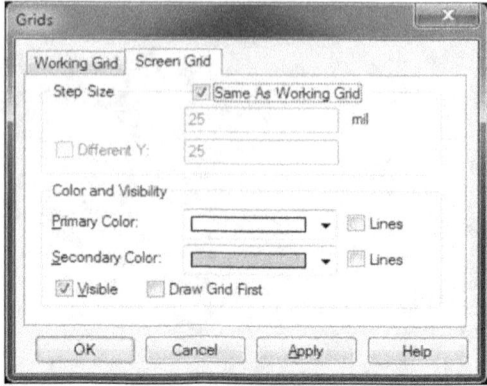

Note: Grid *settings* has a pencil in the icon. Grid *display* does not have a pencil in the icon. Grid Settings set the spacings and Grid Display toggles the grid off and on.

Note: It is sometimes necessary to turn the Grid Display of then back on to display the grid.

Now the display is set so that it is .025" between horizontal dots. Vertical dots have the same spacing. The primary dots (the bright ones) are spaced at 10 grid spacings or 0.100" with the settings above.

Select the line drawing tool from the menu.

Beginning from the top (arbitrary) draw a line two dots down and two dots to the right. Beginning at the end of the previous line draw a line four dots down and four dots to the left. Beginning at the end of the previous line draw a line four dots down and four dots to the right. Beginning at the end of the previous line draw a line four dots down and four dots to the left. Beginning at the end of the previous line draw a line two dots down and two dots to the right.

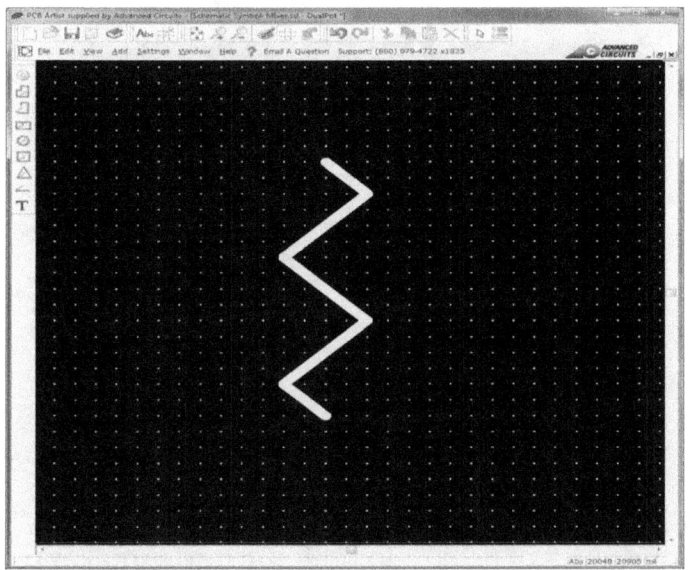

It is now necessary to draw some lines to the connection points. To match the rest of the schematic drawing these connection points should be a multiple of 0.2" apart. Draw a line from the top line up four dots. Draw a line from the top of the previous line four dots to the left.

From the bottom of the drawing draw a line four dots down. Draw another line from the end of the previous line four dots to the left.

We now need an arrow to denote that the resistor is adjustable. Beginning twelve dots above the bottom line (or twelve dots below the top line) and draw an arrow that ends sixteen dots to the right of the left end of the left most lines.

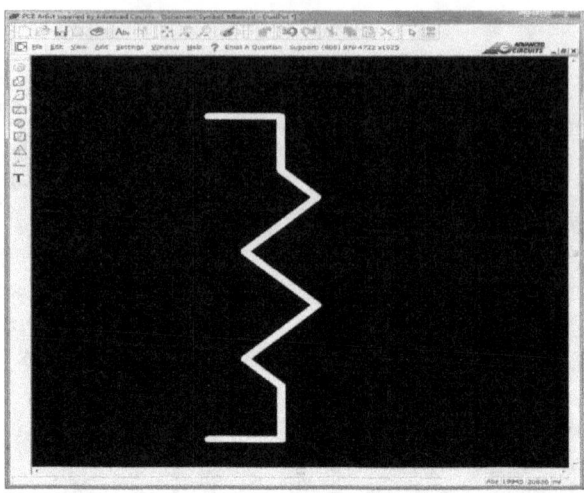

This gives us a variable resistor that has connecting points that are multiples of 0.200" in both the horizontal and vertical axis.

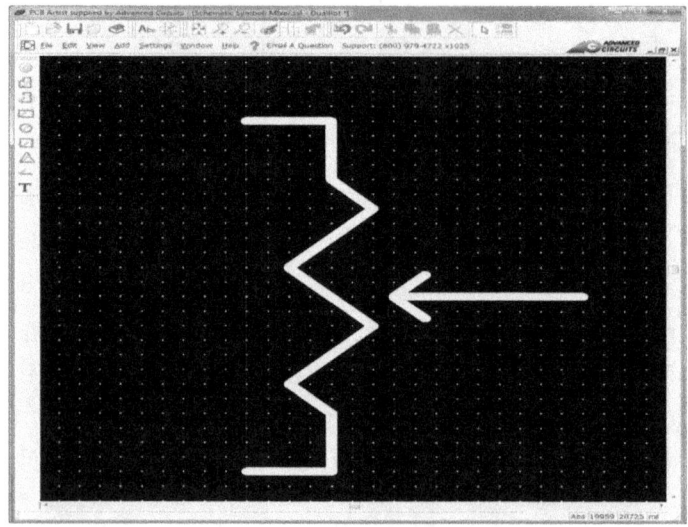

You now have a single potentiometer drawn. Now you can draw a box around all the component you just drew press Cnt-C and then Cnt-V to create another copy of the potentiometer. Position the new copy directly below (or above) the first copy and press enter (or left mouse button).

You can now move the connection points to the ends of the construction lines as shown here

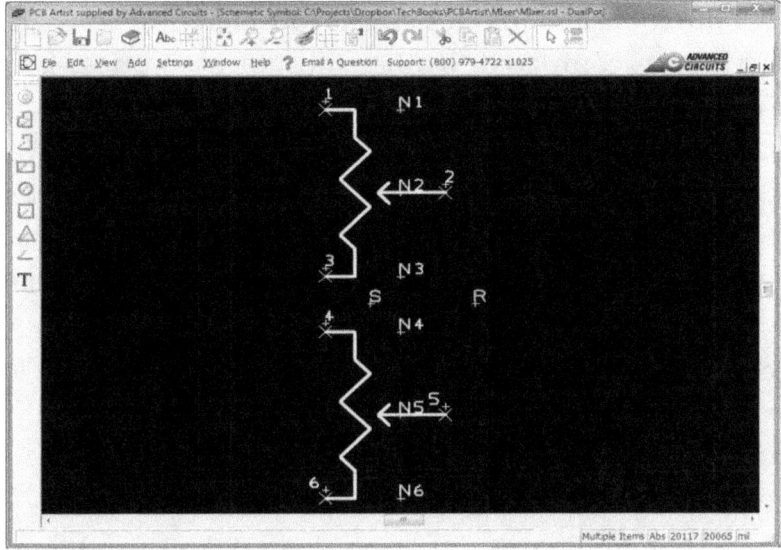

The thru-hole schematic symbol for our dual (stereo) pot is completed.

The grid dots are not visible on this drawing. The visibility of the dots is largely a function of the display card and monitor. The dots show up well on my laptop.

PCB Symbols

A conductor is a material that has low electrical resistance to current flow as compared to other substances. Copper is a good conductor and fiberglass is a poor conductor. A circuit is a continuous conductive path from one terminal of a power source, such as a battery, to the other terminal of that voltage source. Early circuits consisted of physical wires connected between components placed there by hand soldering. That process was tedious and time consuming.

If copper is placed on fiberglass the copper provides a low resistance to current flow and the fiberglass provides mechanical stability and electrical separation (isolation) between copper paths (traces.)

Printed Circuit Boards are planar in that they are shaped with a fixed thickness that is small compared to the height and width of the board. Board thickness is most commonly approximately 0.062." Other common sizes are .032", 092" and .128".

The primary reason for using printed circuit boards is that the circuit is printed much like printing a poster. There are two sided and one sided boards as determined by the number of external planar surfaces that have copper conductors.

Connection density may be improved by adding internal conducting layers. These boards are referred to as multilayer boards. A common multilayer board is a four layer board with two external layers as described above and two internal layers which are often dedicated to power and ground. The power and ground nets usually have the most connections in a circuit.

Experience has shown that connection between the two planar surfaces is best accomplished during the manufacture of the printed circuit board.

One traditional method to manufacture a printed circuit is to glue a layer of copper on each side of a fiberglass resin board. The board is first drilled and seeded[4] to create a conductive surface inside the drilled holes.

Then electrodes are connected to the foil on each side of the board. The connection to the foil is to provide a conductor in the barrel of each hole so that copper electroplated on the walls of the hole. This is done only for holes that need to have conduction from one side of the board to the other. Current can travel from one side of the board via these connections. The plated holes are called vias. The unplated holes are referred to as, interestingly enough, holes.

The after the holes are in place a resist is placed on the copper. The copper will not etch where there is resist on the surface. In a positive process the resist used is not easily washed off....unless exposed to an ultraviolet light. Opaque artwork is placed between the resist and an ultraviolet light. Where the light strikes the resist the resist softens. The resist that was protected from the ultraviolet light by the artwork remains when the resist is washed off resulting in a film that protects the copper under the resist during etching. The boards are then placed into an etching bath where the unprotected copper is removed from the surface of the board. After etching the copper gives conductive paths between component pins through the vias and traces.

---

[4] Seeding is the process of creating a conductive coating in the barrel of a drilled hole so that copper may be plated in the hole. This coating is often copper based.

The hole drilling equipment manufacturer that dominated the early market was Excellon. The company published the interface information to their equipment so that software vendors could write printed circuit board drilling software using their codes. Being the most stable and best documented drilling equipment Excellon dominated the market. Thus a file that can be used to drill the holes in a PCB is most often called an Excellon file.

In a like manner the company that dominated the artwork creation market was Gerber. Thus the files that create artwork for manufacturing the copper layers on a board are most often called Gerber files.

Printed circuit artwork software generates the Excellon and Gerber files as well with many packages providing additional files for other uses such as programming automated manufacturing equipment and mechanical drawing software.

PCB Artist creates the files that can be sent to Advanced Circuits to create a printed circuit board. The PCB symbol editor provides the link between the human readable schematic editor and the Excellon / Gerber type files.

Originally, nearly all components were connected by means of plated thru holes in the PCB. Now most components sold do not require holes as they are surface mount components that are connected to pads on the surface of the printed circuit board.

We will create one surface mount PCB component (opamp) and one through hole PCB component for this document.

Creating Surface Mount PCB Symbols

Surface mount components mount flush to the printed circuit board and solder to exposed surfaces referred to as pads. When manufacturing surface mount board products the PCB is thoroughly cleaned. Then solder paste is deposited on the surface of the PCB usually by means of a stencil.

Solder paste is a mixture of solder, rosin and other proprietary constituants. The solder paste is placed on the PCB and components are placed on the solder paste. The rosin melts first and cleans the surface of oxides. Subsequently, as the temperature increases the solder melts and wets both the pad and component pin.

Mechanical alignment is much easier than it first seems. For small component misalignment surface tension of the solder pulls the pins of the component into place as it floats on the solder pool(s.) It is truly amazing to watch a misalligned component 'float' into correct position as the solder melts.

One popular manufacturing process is to screen the board with solder paste, place the surface mount components, heat the board to solder the surface mount components. Then place the through hole components and solder them. Lastly the 'hand placed' components are added to the assembly.

Please refer to the attachment SO-8 Package for mechanical data of the specific package we will use.

To create a new unique library you create click on the folders tab in Library Manager. Assuming you have followed the tutorial for creating schematic symbols you will have the paths to folders set.

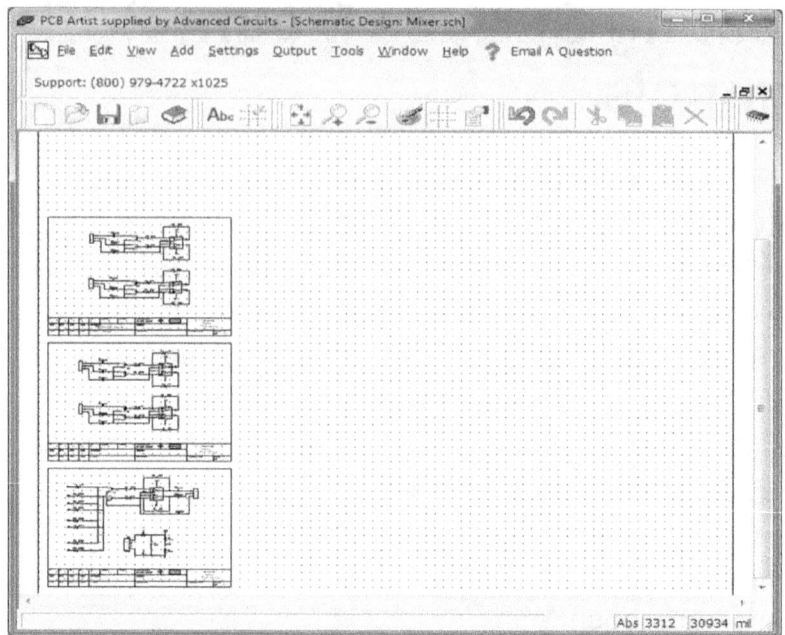

You than click on add which will give you a Browse box. You can then either type in the folder or click browse. In this case I'm using the mixer project folder as we did in the schematic capture section of this document.

Generally you would want to choose a single location to which all your project would point. I often use

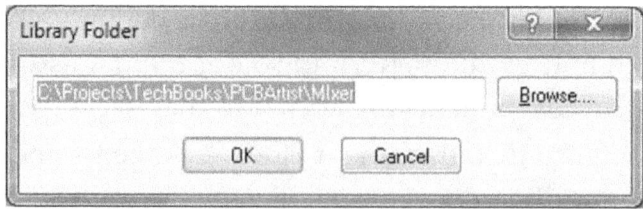

c:\projects\library\ for my common libraries.

Then you would select the Schematic Symbols tab in the File → libraries window.

Creating a New PCB Symbol Library

We will create a new library by clicking the New Lib... box. I've put the new file in the Mixer directory.

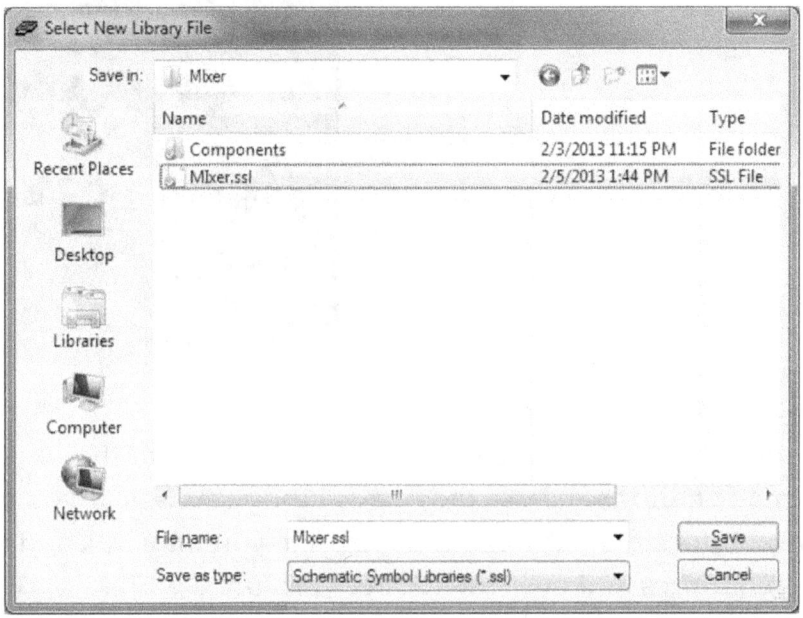

For this project I'm putting everything in the Mixer library. I personally use a number of libraries I've developed with names like HuntSemi.cml for my semiconductors, HuntCapacitorTH.cml for thru-hole capacitors and so forth.

Again I caution you about putting updated components in libraries with the names given by PCB Artist. They will be over written upon upgrades. They caution you about this also, but it bears reminding so that you won't loose valuable time and work.

PCB Symbols

Now we will invoke the PCB Symbols Wizard giving us the

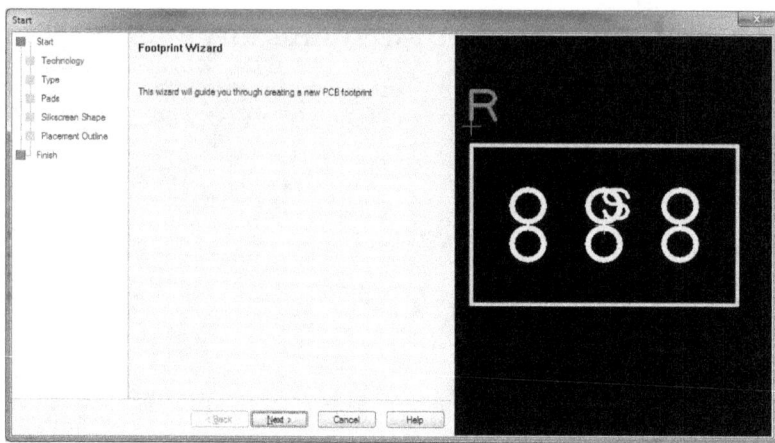

Footprint Wizard screen.

Press next to get to the technology screen. As in the schematic editor we'll use the English thou settings. If you are more comfortable with millimeters units you may mix and match libraries and even switch between units while constructing components and symbols.

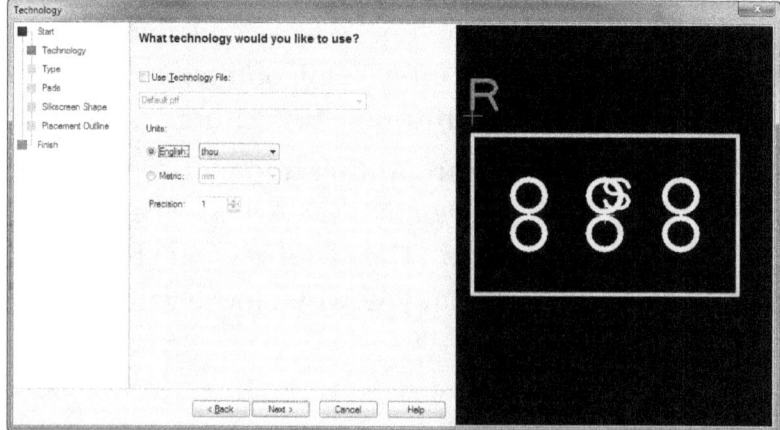

Press next to get to the Type screen. Our part is best defined as an SOIC. The primary differences in the types of foot prints are through hole technology vs. surface mount and

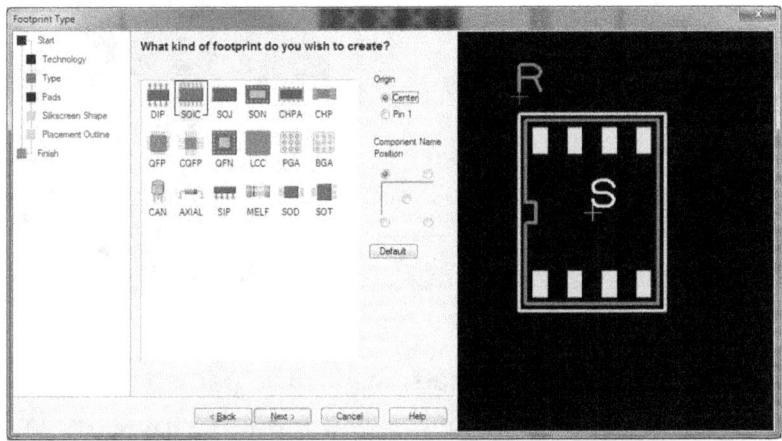

the method of number the connections.

Press next to get to the Pads section. The Pin Counts is 8 for this device. If the component has mounting there will be more pads than connections. The pad style is rectangular

Style: is not selected here. However, if you know that a certain pad style is appropriate for this symbol you can select it here. Usually you would want to do each on a case per case basis.

The sections e: and E: refer to the spacing between pins and the total width of the conductive portion of the package. From the data sheet we learn that e: is .050″ or 50 since the measurement system is thousandths of an inch. From the data sheet we get 236 thou (rounded up) for the total conductor width from the outside end of pin one to the outside end of pin 8.

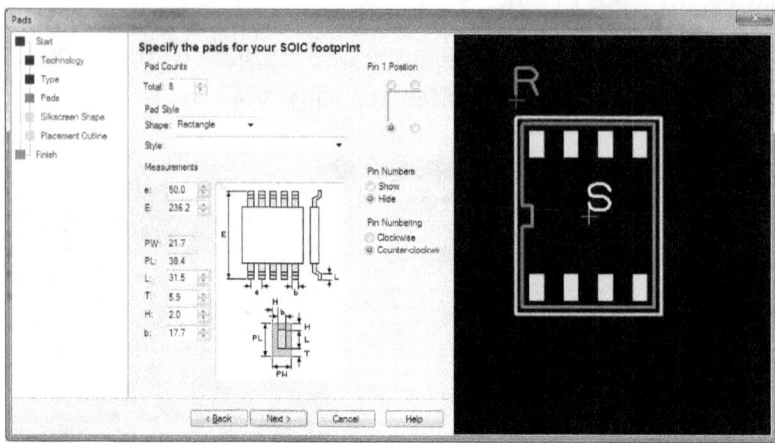

Note that there are four settings for the pad shape. L: and b: set the conductive portion of the pad model. However we also have a PW and PL which represent the size of the solder mask.

Solder Mask

A solder mask covers the printed circuit board except where solder is to be placed. The purpose of the solder mask is to protect the PCB where there will not be any solder. Copper tarnishes quickly but the tin in solder is adequately inert that it was once used in the manufacture of cans for food products.

Soldermask is a printing operation, traditionally a silk screening operation, and the accuracies are such that one must allow a margin of about .005″ around any solder locations otherwise the mask will cover the area to be soldered and corrupt the soldering process.

The mask will also raise that leg of the component resulting in soldering difficulty since this will raise other legs of a component, possibly prevent soldering all together. So we give .005" for T: and H: so that the solder mask clearance is sufficient.

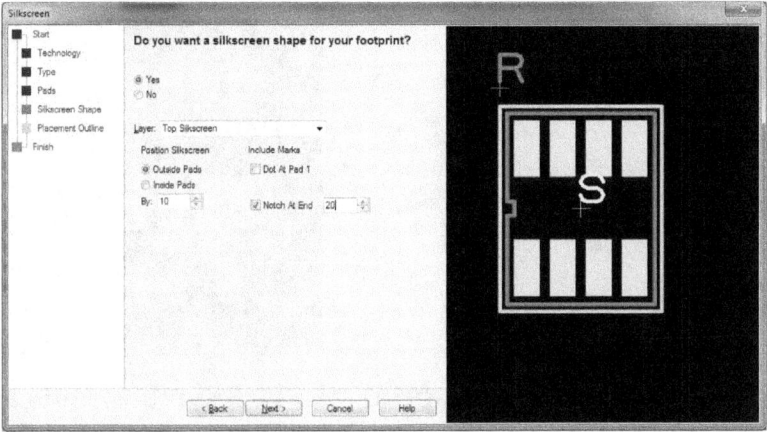

Press next to get to the Silkscreen shape. The silkscreen is printed on the PCB as the last step in the PCB manufacturing process. The primary purpose of the silk screen is to give users an indication of component designations. Other uses are safety information, manufacturer's name, Revision level and similar information.

It is important that silk screen not cover bare copper as it will cause connection problems. Creating a silkscreen is a good idea even if it might not be used at this time. Notice the Layer: pull down menu the placement information is typically on both the silk screen layer and the documentation layer. The options shown are adequate for this PCB model.

Press next for the placement shape for this model. The placement model is slightly larger than the silkscreen model. This gives the PCB designer and indication as to how closely components may be placed.

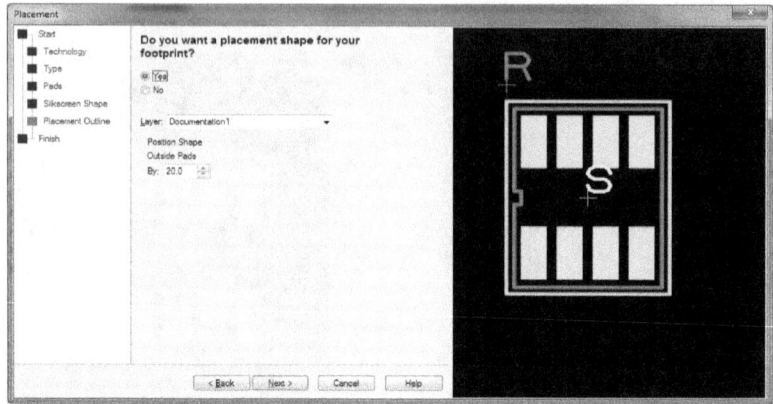

We have finished the footprint (aka PCB symbol) and can save it to the library. You type in the name you want to use for this component. A generic name is appropriate for parts that are expected to be reused. The SO8 package is in common usage so I have given it a generic name. In the case of the thru-hole component created elsewhere in this book the part is named for that component and only that component with a very long name.

Press next to Finish the symbol by entering the symbol name and selecting the library to which it will be stored.

A note of caution. Don't take manufacturers proclamation of a component's PCB symbol without at least a cursory examination. Many times different manufacturers will proclaim they utilize a particular layout and, for whatever reason, their implementation is slightly different than other vendors. This is especially in the case of surface mount components.

The same is true with other libraries. Be sure to double check that components meet your own requirements.

Creating Thru-hole PCB Symbols

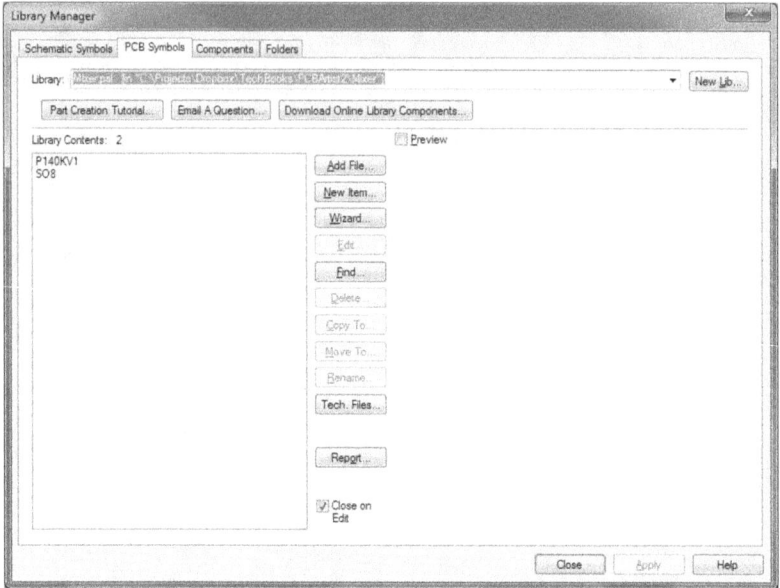

A thru-hole PCB symbol is created in a very similar manner as a surface mount component. However, we will start from scratch since you may never have a reason to create a surface mount component . We begin by selecting File->Library or typing ^L (Control L) to open up the Library Manager window.

Ignore the component above is the SO-8 package. It is the residue of the last symbol created. This is the symbol we created for the chapter on Surface Mount PCB Symbol Creation chapter of this book.

We will create the component using the Wizard so we press the Wizard button giving the Footprint Wizard dialog box. To the left is the familiar progress screen. We are at the start of creation.

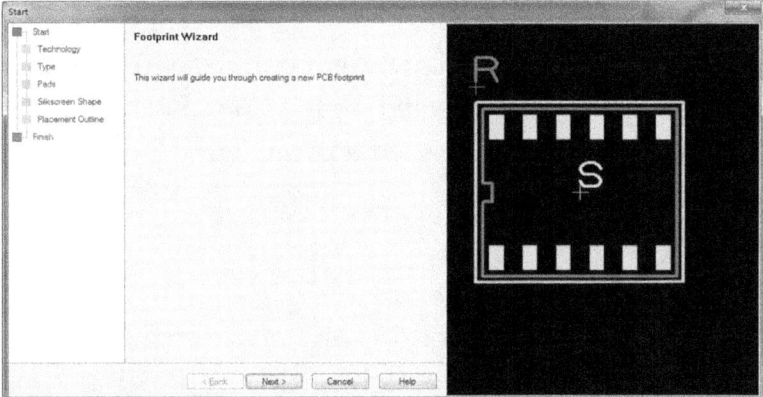

Select Next.

Technology:

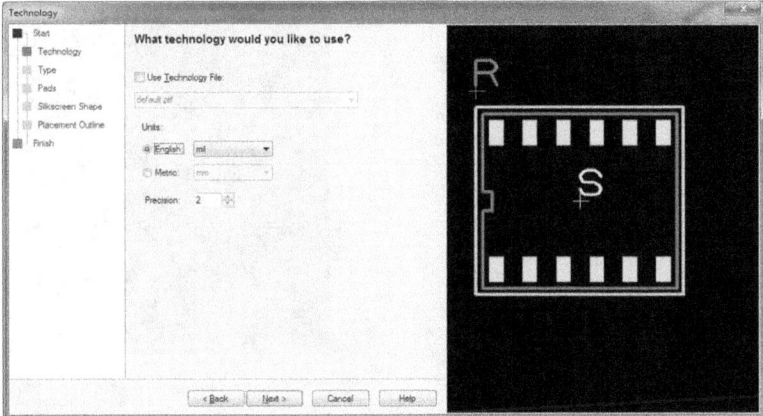

Units:

We select the radio button for English and select thou from the pull-down screen. A common mistake here is to select inches. When generating the model you want to enter 0.1" in inches you would type in .1 using inches or 100 using thou.

Precision:

The precision section is dependent on the setting of the Units section. If thou is selected a precision of 0 or 1 is appropriate. If inches is selected a precision of 3 or 4 is appropriate. Select next.

Footprint

Our part doesn't correspond to any of the available selections available. However either DIP or SIP will give us a thru-hole model that we can alter for our needs.

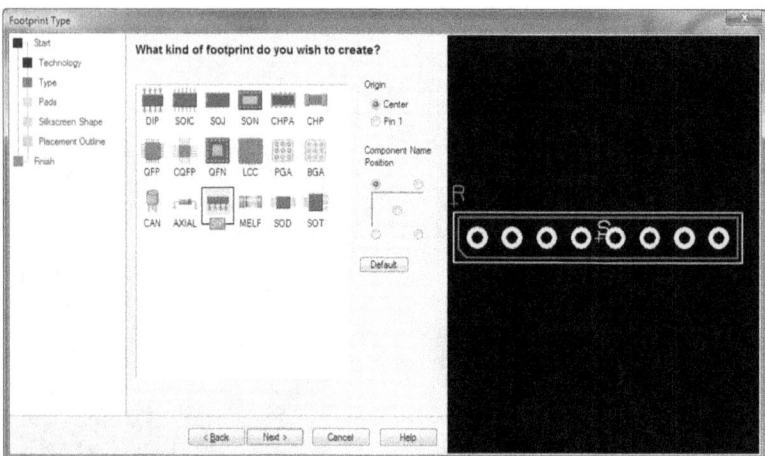

I selected SIP. The SIP footprint is a number of through hole pads arranged in a line. A silk screen outline and a documentation outline are also provided and you are given a chance to modify those outlines as well as the pad sizes and drill holes.

There are some things you need to know if you were to select the DIP (upper left) foot print. You are given the option of selecting a center or pin one origin. I select the Center origin for most parts so that I may easily spin the part during placement and examine the effects. The trade-off is that some manufacturing packages (for example to create pick and place design files) need pin 1 location. You should also know that you can change the origin location at any time.

Component Name Position is also easily moved later. I usually accept the default.

Pads

The pads screen sets the number and shape of the pads to be inserted by the footprint wizard. The PCB symbol has 6 active (connected internally) pins and three pins that are not functionally connected. You should also notice that layout of the pins is unusual, though for a good reason.

I have modified the manufacturer's drawing and appended it to the end of this book as P140KV1 Through Hole Package. If you look at the lower left corner of that page there is a recommended PCB layout. I've converted the original document's metric measurements into Imperial units (inches.)

This device will be the volume control for two channels of a stereo signal. Thus, the part is two potentiometers controlled by a single shaft. This gives reasonable tracking of the two channels of a stereo signal when the knob is turned. It is most convenient for the manufacturer if the low (arbitrary designation) side of both pots are mechanically on the same side of the part. If we were to number the pins 1, 2, 3, 4, 5, 6 and 7 from left to right then pins 1, 4 and 6 are one potentiometer and 2, 3 and 5 are the other potentiometer. Pin seven is a clone of the first six pins with the same spacing.

The other two pins (we'll call them eight and nine) are there to physically mount the potentiometer.

The total pin count for the PCB symbol is nine pins and six pins for the schematic symbol.

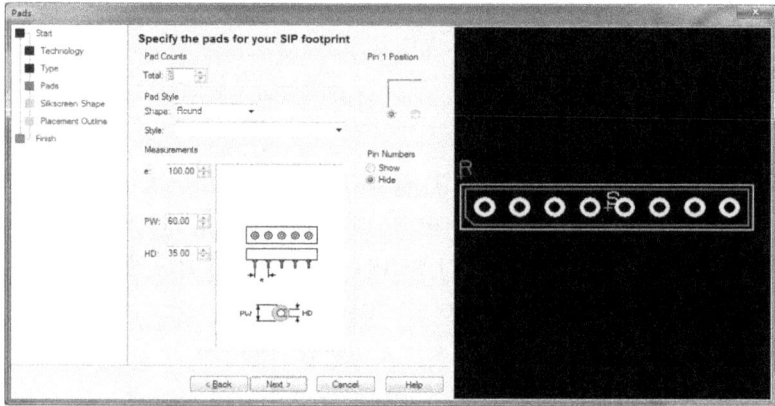

Pad Counts

As explained above we select a total of 9 pins.

Pad Style

The majority of the pads are round and .039" in diameter. I personally like to give a few thousandths of extra clearance so I'm selecting round pads and .042" holes.

Measurements

The symbol e is the spacing between the pads. The spacing between pads is 0.0787" so we set e to 78.7 thou. Remember that we selected thou as the default unit of measure.

PW

This stands for Pad Width. Pad width is primarily determined by HD (hole diameter) below since the minimum pad width must be greater than the hole diameter to provide for an adequate annular ring. The annular ring is the printed circuit foil that is left after the pad is drilled and the circuit is etched. So if we have an pad of .062" and a hole of .042" the annular ring is .010 inch on each side of the hole and the annular ring is .010" wide. (Pad diameter – drill hole diameter.) Among other things this assures that there is copper all around the hole after drilling after all manufacturing tolerances are taken into account.

From the drawing we know the hole size is 0.042" adding .015" gives us a 0.0057" annular ring. This is adequate for signal tracks. The annual ring should be half the standard power track width when power traces are routed to a thru-hole pad.

There is some disagreement in the industry regarding the annular ring for power thru-holes. When a thru-hole pin is soldered the additional cross section of the solder contributes to the current path improving current handling. Further, track width is most often determined by the temperature rise of the foil at maximum current during operation and the pin in the hole increases dissipation. However, the pin may not be fully soldered and component heat might be contributing to the foil temperature. I prefer the conservative approach when possible.

HD

This is the outer diameter of the pin that will be inserted into the pad. We are using a hole diameter as determined by the datasheet.

An additional note regarding hole size. A round pin requires a hole about 0.005" larger than the maximum pin size just to allow for manufacturing tolerances. A clearance of 0.007" or more allows solder to easily contact the length of the plated thru hole barrel. A 0.025" pin (such as standard ribbon cable headers or wire wrap sockets) requires a .043" hole to accommodate the maximum width between opposing corners of the square pin plus any soldering clearances.

Silkscreen

There are a number of things to remember about the silkscreen. Silkscreen printing should never be allowed over a pad or hole, only over traces or open board. Most board positioning is predicated on the silkscreen shapes. To this end the silkscreen should be as small as possible (to not interfere with other components or via holes) and as large as necessary (to provide for adequate spacing between components.)

I've never encountered a situation where a silkscreen was not warranted for a component. So we select the yes radio button.

Layer

I always select the Top Silkscreen layer. It is possible to create a bottom silkscreen if the footprint is to be mounted on the bottom side of the board but keep in mind that any lettering will be reversed and bottom side silkscreen is rarely ordered.

Thankfully, PCB Artist displays the bottom side artwork reversed as it would be seen from the component side of the board.

Position Silkscreen

You may notice that there is no alternative to Outside Pads and that the radio button cannot be deselected.

I usually place components on a .025" grid and use the 25 option.

Include Marks

In this case it does not matter as we will be modifying the silkscreen, but normally you would select both Dot at Pad 1 and Miter Corner.

Select Next.

Placement Outline.

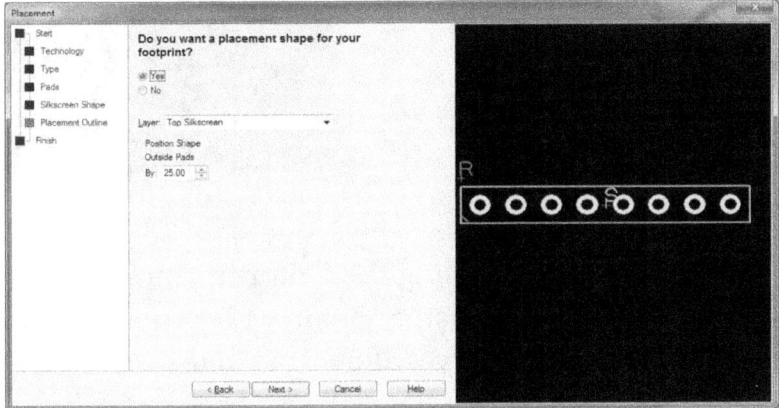

I usually use the Silkscreen outline for placement. Because it is not a part of the board manufacturing process the documentation1 layer is often used to create a printable representation of the board that includes items such as notes, extra information and information can be placed anywhere on the surface. For this reason I reproduce the silkscreen values in the placement screen.

Finish

Finally the PCB symbol is ready to be created.

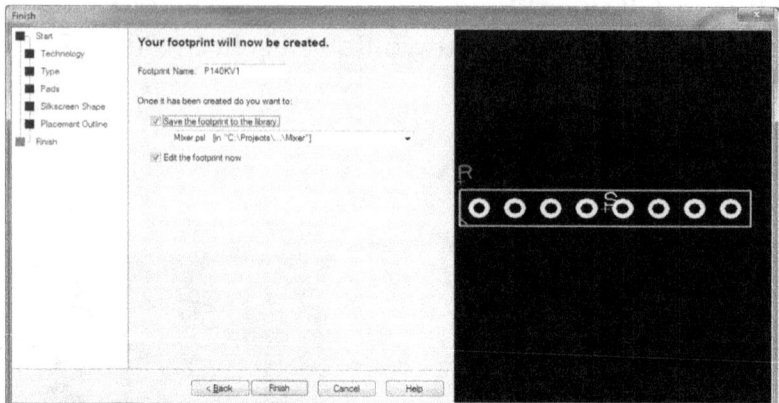

Footprint Name:

If you have deferred naming the footprint you place the name here. I selected this name, P140KV1 because it uniquely describes the footprint, but yet allows me to use the footprint for component symbols that have other tapers and values. There are a number of models of potentiometer that share this footprint and this symbol can be used for any of them.

Save the footprint to the library:

I always check this so that I have a fallback copy just in case I make a misteak during editing and we will be doing a lot of editing.

Edit the footprint now:

Many of the library parts created do not need any further editing and this is not checked. But that wouldn't be very illuminating would it?

You might have a screen like the one to the screen below after the wizard finishes. In this case, and in any case you can press the 'view all' button.

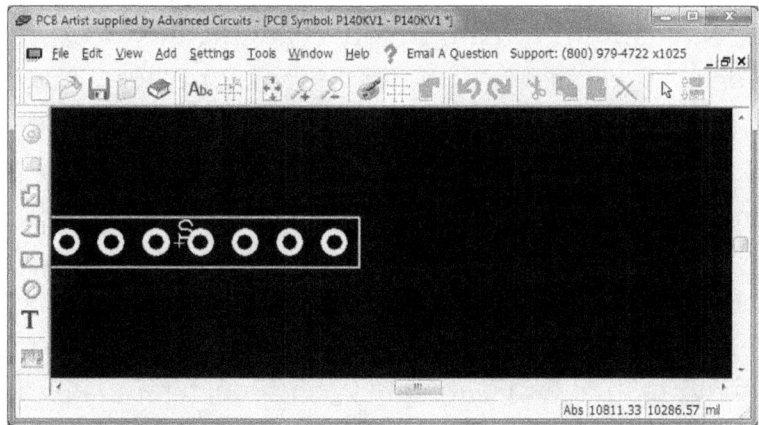

The symbol will then become scaled and centered on the computer's screen. This button is useful in nearly any situation that might arise where you have lost track of the item you are editing.

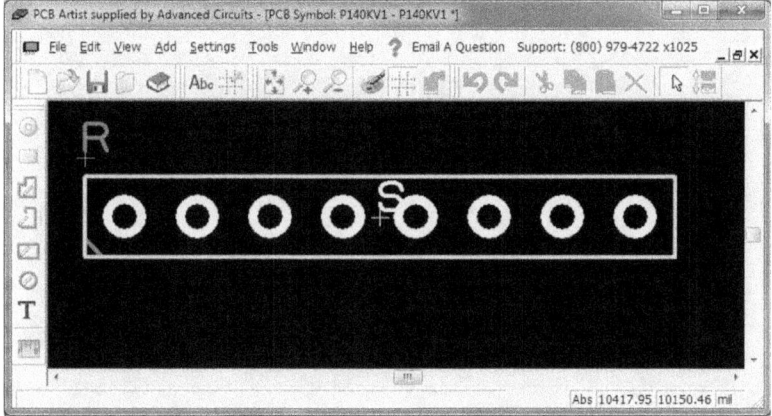

You now have a centered symbol ready for editing. You may have noticed that some items such as the silkscreen and placement outline change colors from time to time. There is a line on the lower left corner of the symbol to the right that is red. If you click on that line the symbols outline will turn red with a green corner. You are simply changing the focus from silkscreen (in this case) to documentation1 (or placement.)

The first thing we want to do examine the schematic capture symbol and compare it to the PCB symbol and compare those to the mechanical drawing (located in the rear of this book.) The schematic symbol has the pin numbers placed neatly from the top to the bottom. The PCB symbol currently has the pins arranged in 1, 2, 3 order. However the mechanical drawing shows that pins 1, 4 and 6 are one potentiometer and 2, 3 and 5 are the other potentiometer.

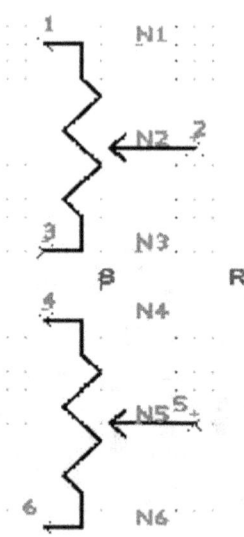

We could move the PCB symbol pins around to correspond to the mechanical drawing, but there is another method I'm going to show you because it is more versatile.

Editing the PCB symbol

First let's get the lines out of the way by clicking on them and dragging. Then click on the two right most pads and move them above and to the right and left. This should give you a screen similar to the one on the right. There is nothing critical about any of these placements as long as you don't move pads 1 through 7.

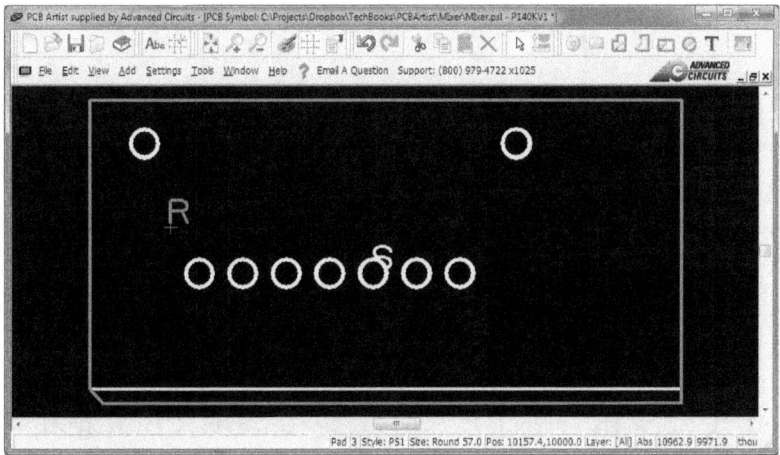

Note that the S indicates the symbol center.

Examination of the recommended PC board layout in the lower left corner of the mechanical drawing will show that the drawing is dimensioned from the center of the fourth pad. We'll use a feature of the settings menu to make placement of the symbol details easier. I promise it will be worth the trouble.

First click on the fourth pad from the left. It will turn white as it is selected. Then right click bringing up an on-screen menu. Then select properties at the bottom of the menu and it will give you a screen like the one to the right. You must take note of the two position entries in the screen to the right. In the example below, these are 10236.1 and 10000.0 your results will probably vary. The left number is the x coordinate and the right number is the y coordinate. You might want to write these values down on a notepad. In our case the numbers are 10236.1 and 10000.0. Your values may vary. You can now select the cancel button.

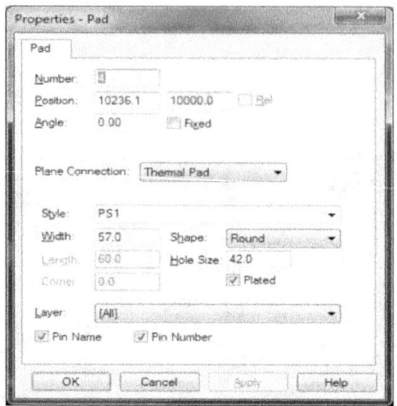

Type AltS → C to get the Coordinates screen as shown on the left. Check the check box for Use Relative Coordinates and fill in the values that you got from the location of pad four in the previous step.

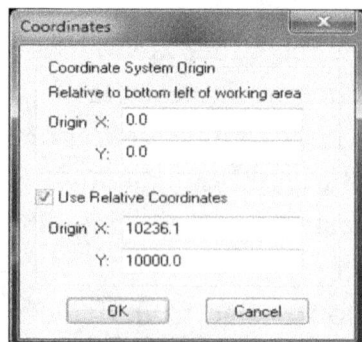

You should get a screen similar to the following screen. The significant item is the x shaped in the middle of the fourth pad from the left. If you right click on the pad → left click and select properties you should get a position pair of 0 and 0. This is good because you will be moving the two upper pads in relationship to pad four as is denoted in the mechanical drawing. Note that on the layout portion of the mechanical drawing the two upper pads are 0.622″ apart. Then note that in the upper right portion of the mechanical drawing the pads are symmetrical to pin 4. Also note that both pads are 0.130″ above pad 4.

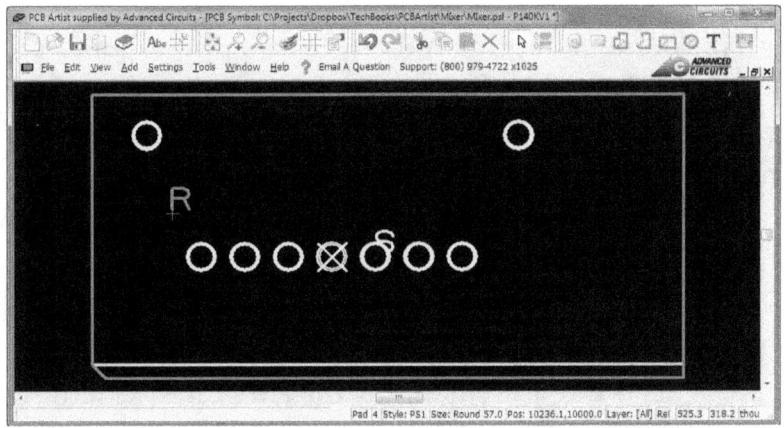

We are going to place these holes by means of direct entry. The left pad is half of 0.622″ to the left of pad 4 so it's x coordinate is -311 it is 0.130″ above pad 4 so it's y cordinate is 121. Select the left most upper pad (left click) and right click and select properties. Type -311 into the first (x) box and 121 into the second (y) box. The right most upper pad is to the right and above pad four so its coordinates are 311(not negative) and 121. All the holes are now placed.

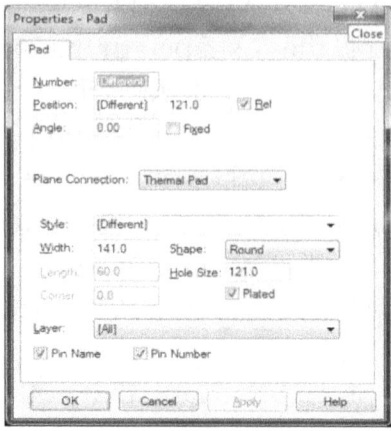

The size of the mounting holes (the two upper holes) are still not the right size. Click on one hole and holding down the control key on the keyboard click the other upper hole. The right click selecting the properties screen for both holes. The recommended hole size is 0.121" so we type 121 into the Hole Size box and for a 0.010" annular ring we type 141 into the Width box. The mounting holes are now placed and sized. Note that even though these are holes the plated box is still checked. This is a requirement for the Advanced Circuit prototype board services.

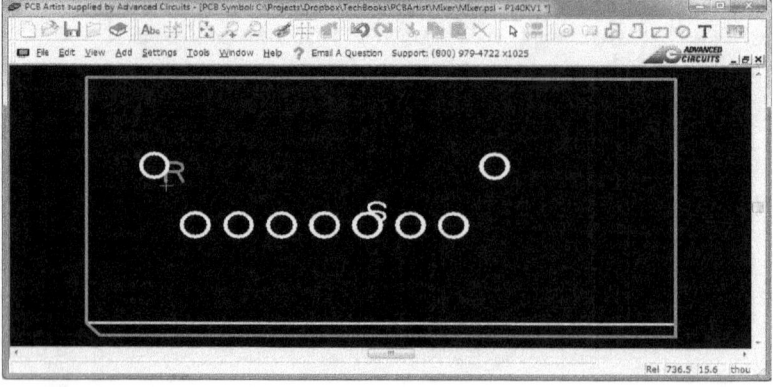

The distance to the back top line of the component is specified as 0.220" which we can set by selecting the top line of the documentation1 outline and dragging down until the right most item in the status box indicates 225. The width is 0.531" so we drag the left side of the documentation1 outline to 0.265" which is the next multiple of the snap setting. The reason we used the documentation1 outline is that it can go across holes and pads.

The Silkscreen

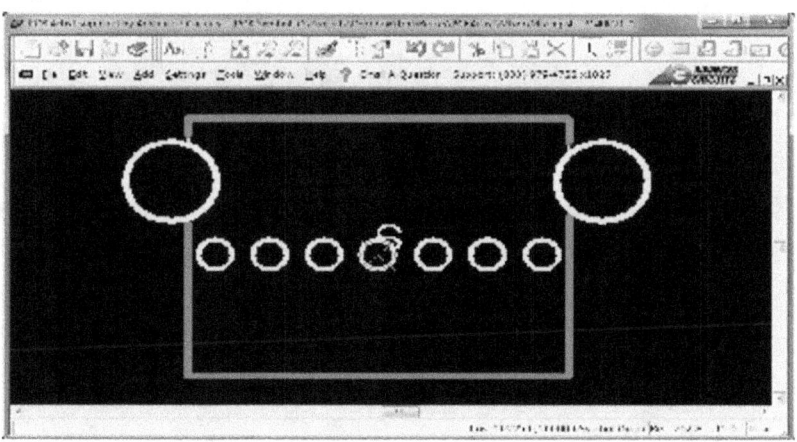 We can now draw the silkscreen outline. Select the Add Open Shape icon (left) to draw open ended lines.

Right click in an open area of the drawing and select Change Layer. Select Top Silkscreen from the pull-down menu and select OK. Click on the green line immediately above the left mounting hole (arbitrary) and begin placing a line directly over the documentation1 outline over to the same point above the other mounting hole.   Right click and select finish here to end the line. Do the same on the bottom portion placing the silkscreen outline over the documentation1 outline.

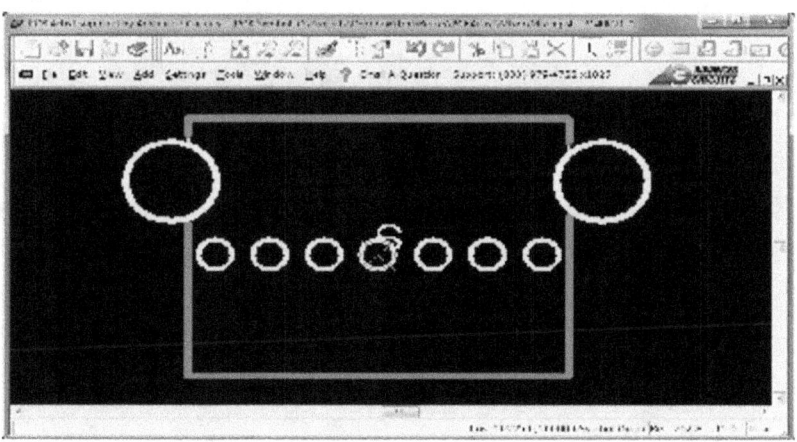

The silkscreen doesn't need to have the shaft and bushing drawn but it would be handy on the documentation1 layer so I added in the same manner as I added the lines to the silkscreen in the previous step.

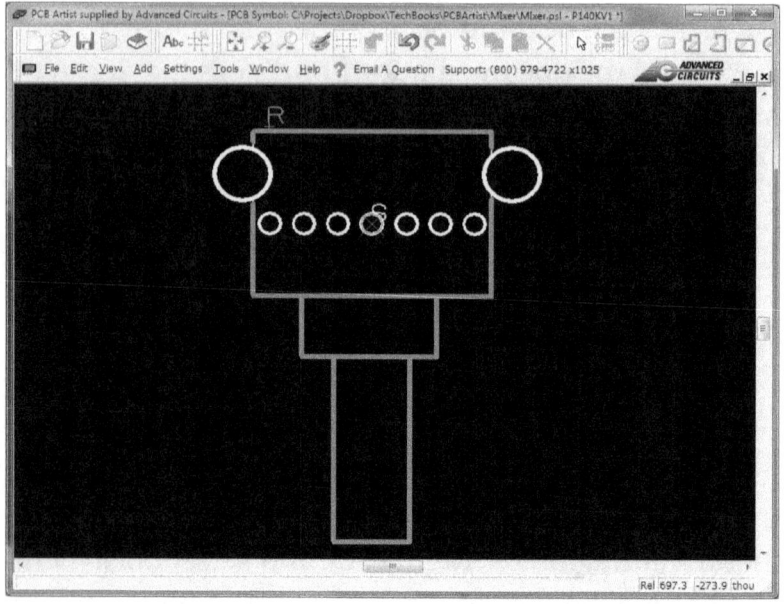

Now select file close. You will get the following dialogue box. Select Yes.

Creating New Components

Many components can be created without creating symbols. This is because components can use the exact same Schematic symbols and PCB Symbols but be different components. An example would be resistors. There is no significant difference between Surface Mount and Thru-hole component creation.

Wizard

Type Cnt-L to start the Library Manager and select the Components tab.

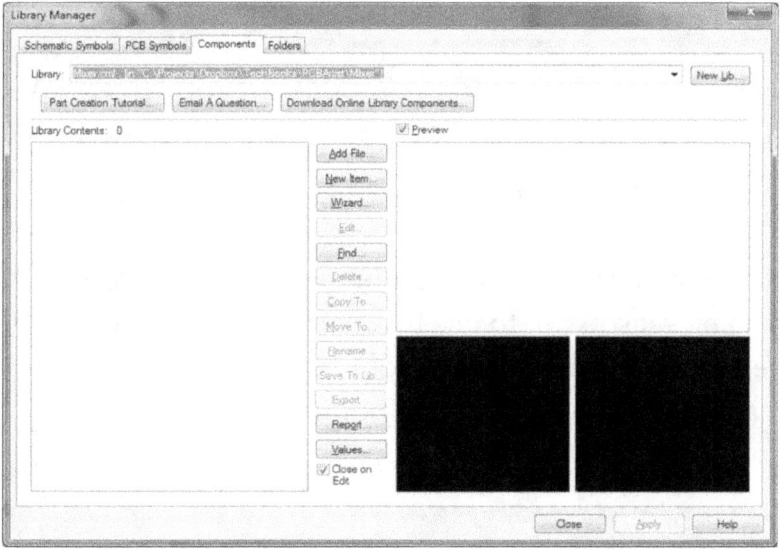

We will select the Wizard button. This gives us the Component Wizard screen.

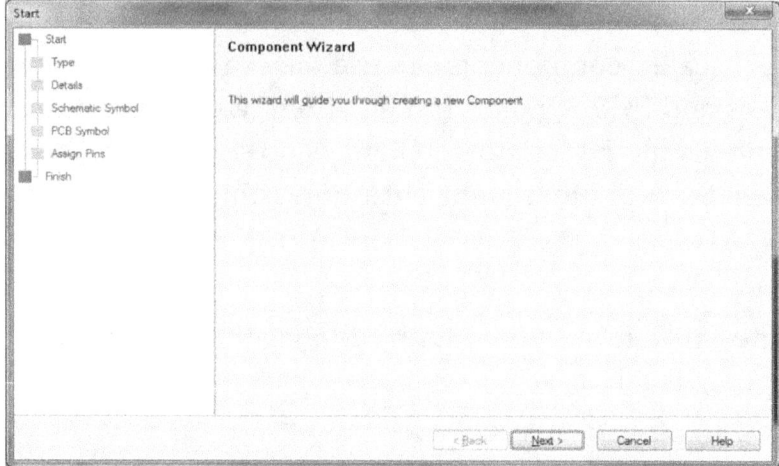

Clicking next gives us the Component Type Screen. We will select Normal Component and select next.

Component Details

The full component part number is P140KV1-F20AR50K. There is no reason not to use this part number since we will not be able to source the part from any other manufacturer and other parts from this series will not work.

The total number of pins for the PCB symbol is 9 even though the Schematic Symbol is only 6 pins.

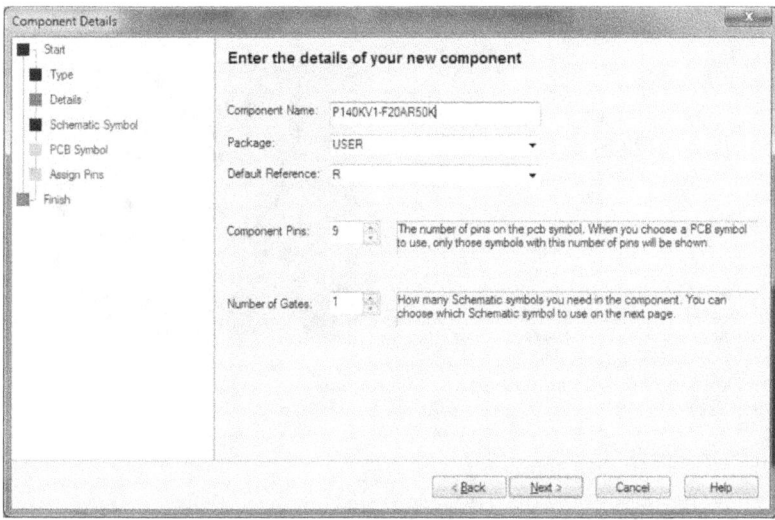

Schematic Symbol Selection

Select next to get the Schematic Symbol selection screen. Assure that Preview is checked.

Select the correct library file, which in our case is Mixer and select Dual Pot.

Note that there is a Find Symbol button. Pressing this will give you a search window. Just remember that if you use pin count in this search window that it is the number of pins on the schematic symbol and NOT the number of pins on the PCB symbol.

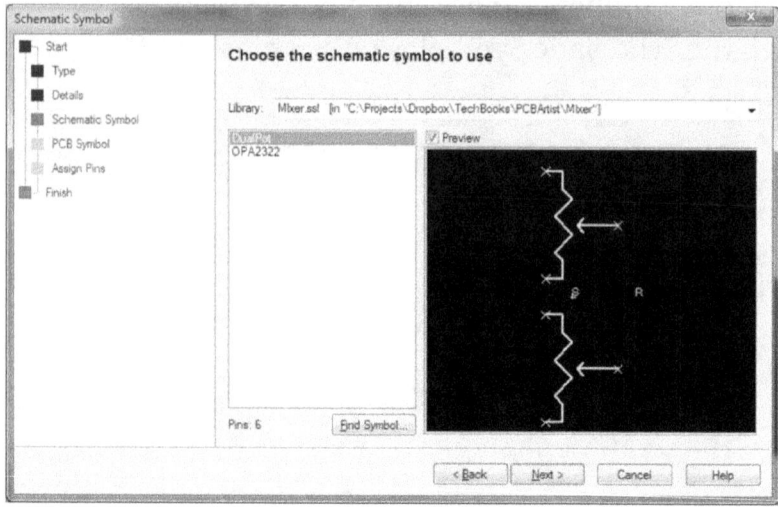

Select next.

PCB Symbol Selection

Assure that the correct library is selected in the Library pull-down menu.

Select the P140KV1 symbol and press next. If you use the find from this page and select pin count as a search qualifier it will require the number of pins in the PCB symbol.

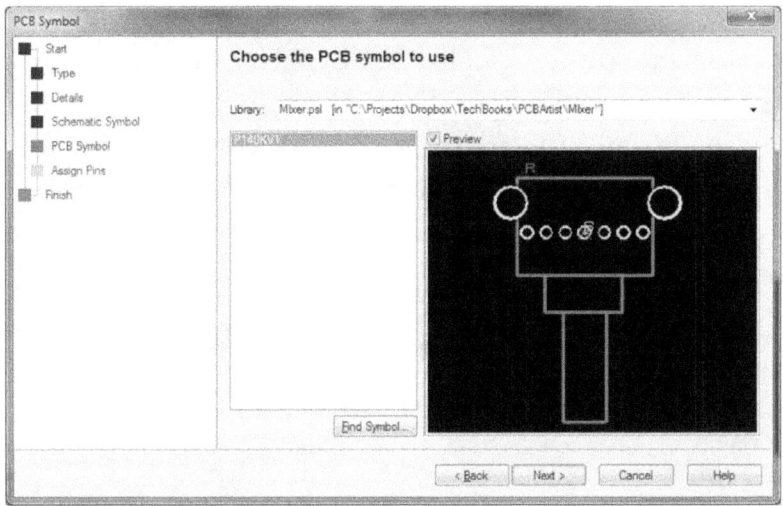

This is a very difficult section. Remember that during symbol definition the pins for the schematic symbol and pins for the PCB symbol did not match up. This is where we put it right. I have arbitrarily designated the upper pot as the Left pot and the lower pot as the Right pot on the schematic symbol. The Left pot on the schematic symbol is pins 1, 2 and three. The Left pot on the PCB symbol is pins 1, 4 and 6. Therefore, to match the Schematic symbol pins to the correct PCB symbol I've typed in 1, 4 and 6 for the top three PCB model pins to match pins 1, 2, and 3 for the respective Schematic symbol pins. I also typed in 2, 3 and 5 for the bottom three PCB model pins to match pins 4, 5, and 6 for the respective Schematic symbol pins 4, 5, and 6.

Pin Assignment

I named the terminals for this example. LCCW and RCCW are the pins corresponding to the wiper direction when the shaft is turned counter clockwise. LCW and RCW are the pins corresponding to the wiper direction when the shaft is turned clockwise. LC and RC are the respective center (wiper) pins. To clarify. When the input shaft is turned clockwise the resistance between LC and LCW decreases while the resistance between LC and LCCW increases.

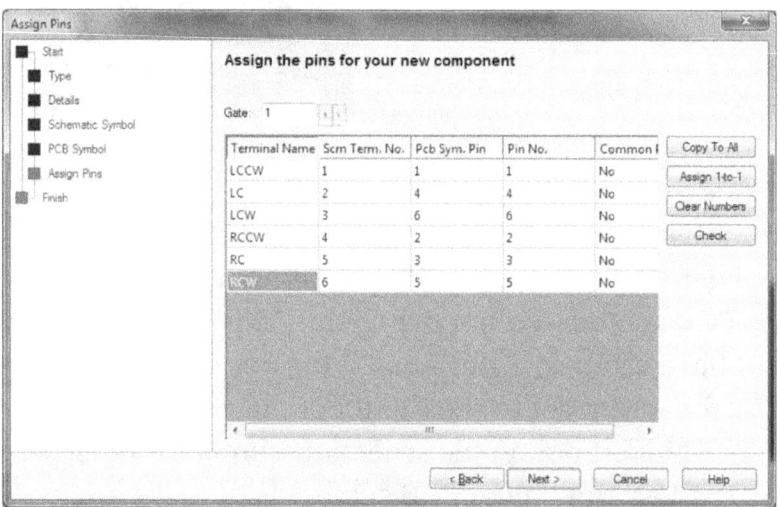

Select next here and assure that the component is being saved to the correct library. You may now check the component by checking the Edit the component now check box.

The component is finished select Finish.

Library Enhancements

Now for one of the "undocumented features" of PCB Artist software. A very accurate, complete and up to date bill of materials can be created by PCB Artist if a bit more information is entered at this stage.

Open the Library Manager screen (Cnt-L) select the component and select Edit. Then press Alt-Enter. This is information that is transferred from the library to the design. The 'trick' is that you can add plenty of information in addition to the information listed here.

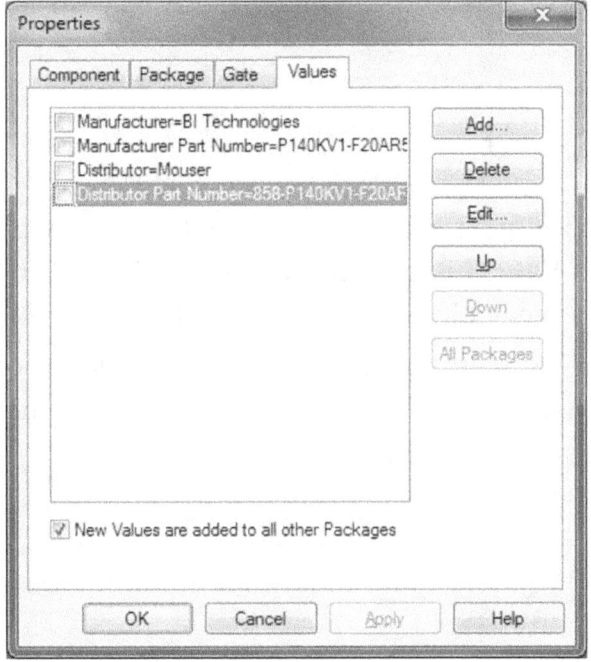

First though let's enter the fields as provided by the software developers. Mouser sells the P140KV1-F20AR50K part under their part number 858-P140KV1-F20AR50K. So let's fill in the fields. Double click (or select Manufacturer and press the Edit button) and enter BI Technologies into the text box. Do the same for each item in turn using the following information:

Manufacturer = BI Technologies

Manufacturer Part Number = P140KV1-F20AR50K

Distributor = Mouser

Distributor Part Number = 858-P140KV1-F20AR50K

Now select the Add button and type Price in the Name: text area and 1.08 in the Value: text area. The Value dialog box should match the illustration below.

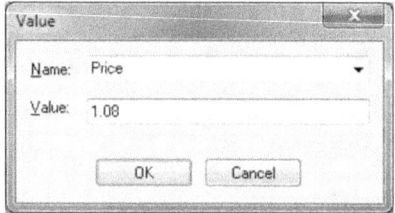

This is a tricky part, there is a Value field and we are going to fill it with Val. This becomes even more confusing (at least to me) when we begin enhancing the Bill of Materials.

We add a Val field which is and set its value to 50K. The result is shown below. We will get back to this when we are ready to create a bill of materials.

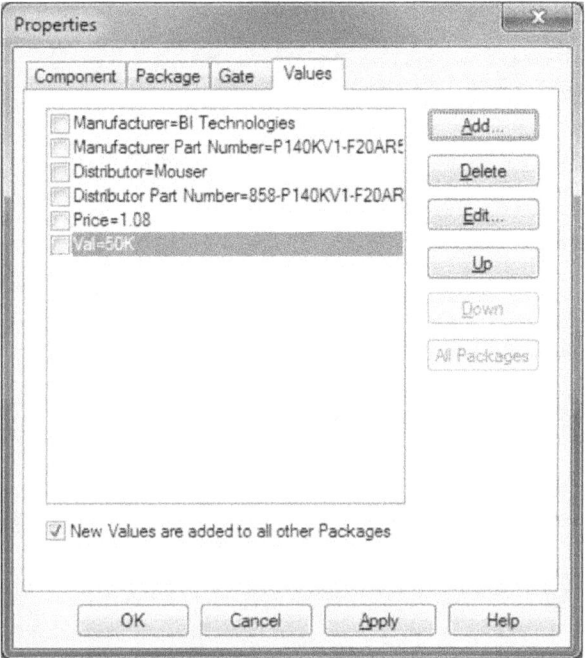

Schematic Reports

Being just barely less than perfect myself I have been known to make misteaks. Error checking is just one of the handy and important features of the reports section of the Schematic Capture portion of PCB Artist.

Dangling Tracks

Type Alt-O then R and select Dangling Tracks from the menu. Select the Run button on the right. This report will tell you if you have a line that goes nowhere. This is the second most common layout error. If you don't find the error until after you order the board it can be devastating. So make the click and save a brick (of money[5].)

---

[5]This brings up similarity between myself and a baseball star. Regarding my jokes I swing, I miss; I swing, I miss; I swing I GET A HIT! If I were to get a .333 laughing average – I'd be an all-star. Practice, practice, practice.

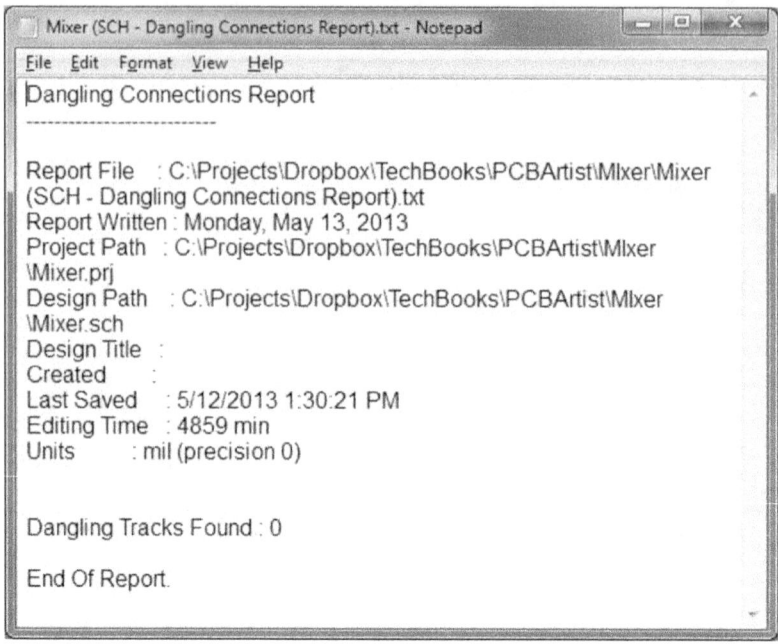

Mixer (SCH - Dangling Connections Report).txt - Notepad

File   Edit   Format   View   Help

Dangling Connections Report

Report File   : C:\Projects\Dropbox\TechBooks\PCBArtist\Mixer\Mixer
(SCH - Dangling Connections Report).txt
Report Written : Monday, May 13, 2013
Project Path   : C:\Projects\Dropbox\TechBooks\PCBArtist\Mixer
\Mixer.prj
Design Path   : C:\Projects\Dropbox\TechBooks\PCBArtist\Mixer
\Mixer.sch
Design Title :
Created   :
Last Saved   : 5/12/2013 1:30:21 PM
Editing Time : 4859 min
Units   : mil (precision 0)

Dangling Tracks Found : 0

End Of Report.

Design Status Report

The next report is the design status report. You will notice that as of this moment I have over 81 hours in this schematic. Creating a schematic for a book requires quite a bit more time than creating a schematic for a product. And I'm going to keep with that story. The schematic symbol you might not recognize is ANSIA. That is the outline drawing for the sheet.

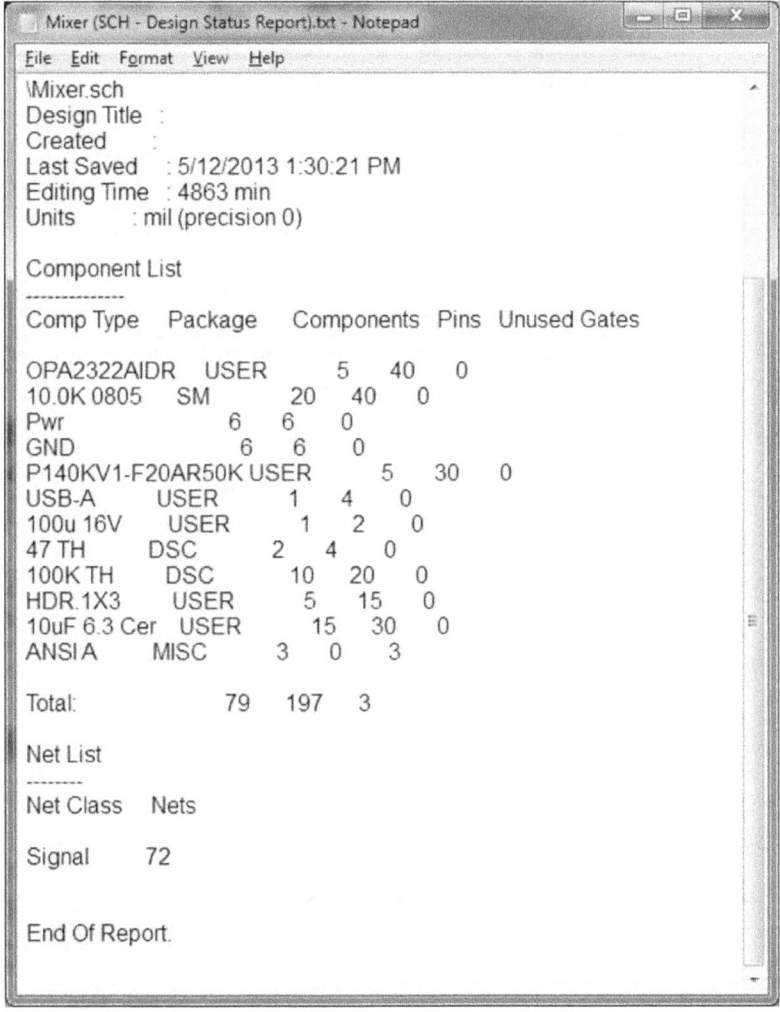

```
Mixer (SCH - Design Status Report).txt - Notepad

File  Edit  Format  View  Help

\Mixer.sch
Design Title  :
Created       :
Last Saved    : 5/12/2013 1:30:21 PM
Editing Time  : 4863 min
Units         : mil (precision 0)

Component List
--------------
Comp Type    Package    Components  Pins  Unused Gates

OPA2322AIDR   USER          5      40     0
10.0K 0805    SM          20       40     0
Pwr                       6        6      0
GND                       6        6      0
P140KV1-F20AR50K USER        5     30     0
USB-A         USER        1        4      0
100u 16V      USER        1        2      0
47 TH         DSC         2        4      0
100K TH       DSC         10       20     0
HDR.1X3       USER        5        15     0
10uF 6.3 Cer  USER          15     30     0
ANSI A        MISC        3        0      3

Total:                    79       197    3

Net List
--------
Net Class    Nets

Signal       72


End Of Report.
```

Generic Netlist

This is often referred to as a 'human readable' net list. When you first select run you get a screen that allows you a number of options. I select all the options.

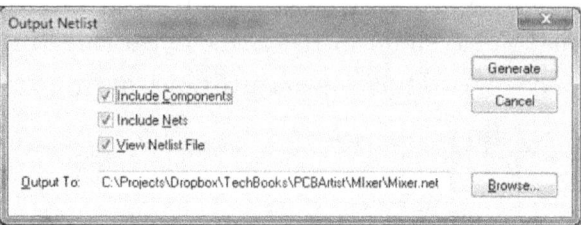

The output is saved to a file as determined by the Output To selection in the previous screen. A portion of the output is illustrated here.

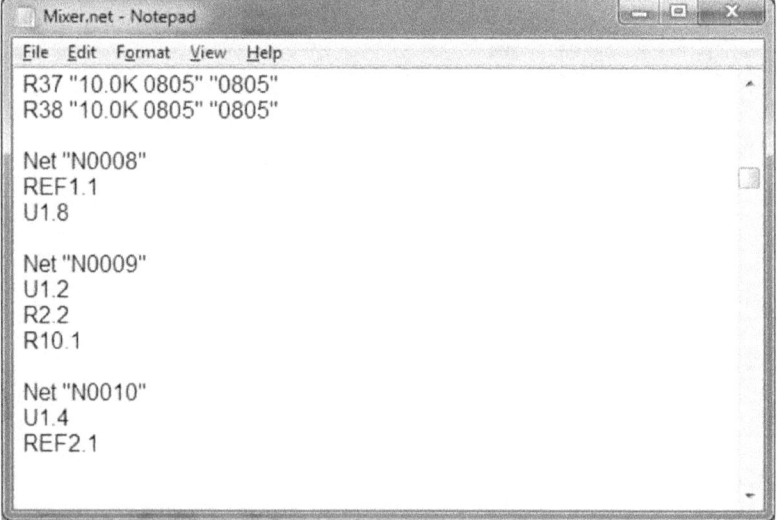

Looking at Net "N0009" you can see that pin 2 of U1 is connected to both pin 2 of R2 and pin 1 of R1. No problem there.

However, if you examine Net "N0008" there are two problems. The first is that it is difficult to determine what REF1.1 represents.

Scrolling up we find that Ref1 is Pwr. Well, that isn't good because we need Pwr to be connected to our supply voltage Vcc. A number of components should be connected to a net called Vcc.

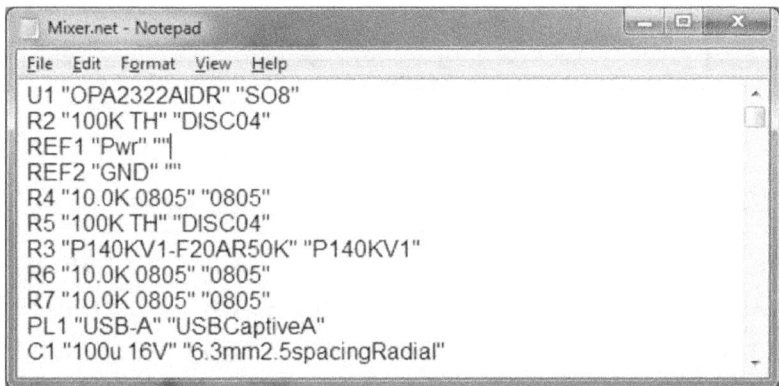

We went to the trouble of removing the default Terminal Name and Net (Class) from the power symbol because there are often multiple grounds in a schematic. For example, a microprocessor with an Analog to Digital Converter (ADC) might have Vss and Analog Vss. The analog Vss would have extra filtering. There would also be a Power Vdd and Analog Vdd. There might also be (and often are) multiple power pins. All of these might have different requirements. For example, you might not want the Analog Vss to be routed on the power layer of a multilayer board.

So, we have to go to every power and ground component and label the line that connects the power and ground. Pain in the clicker finger but it is the best solution. We do this by left clicking on the line to be labeled and then right clicking and selecting Display Net Name and then select Change Net.

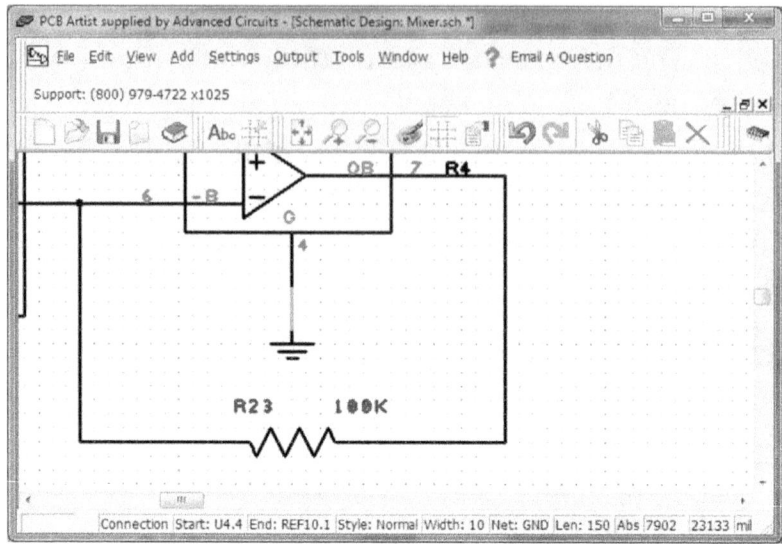

The Change Net screen gives all the existing net names. You can either type in the net name (or a new net name) or select the net name from the list. Sooner or later you will have to decide whether to use the Net Class features of PCB Artist.

Personally, I do not. However, until you understand advanced power and ground routing you should use the Class Name features. It will greatly simplify multilayer boards if you have only one power and ground.

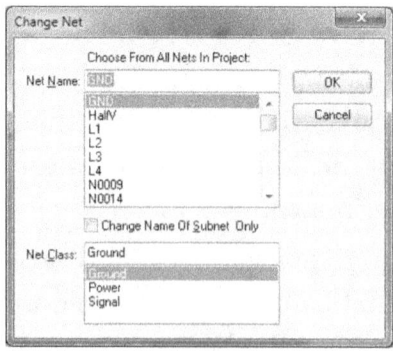

In this simple design there is only GND and Vcc. I have indicated that the board can be powered by a USB connection and called out a captive USB A connector in the bill of materials, but I also plan to operate with batteries.

If we now look at the net list below the Net GND has a number of connections including REF's as does the VCC Net.

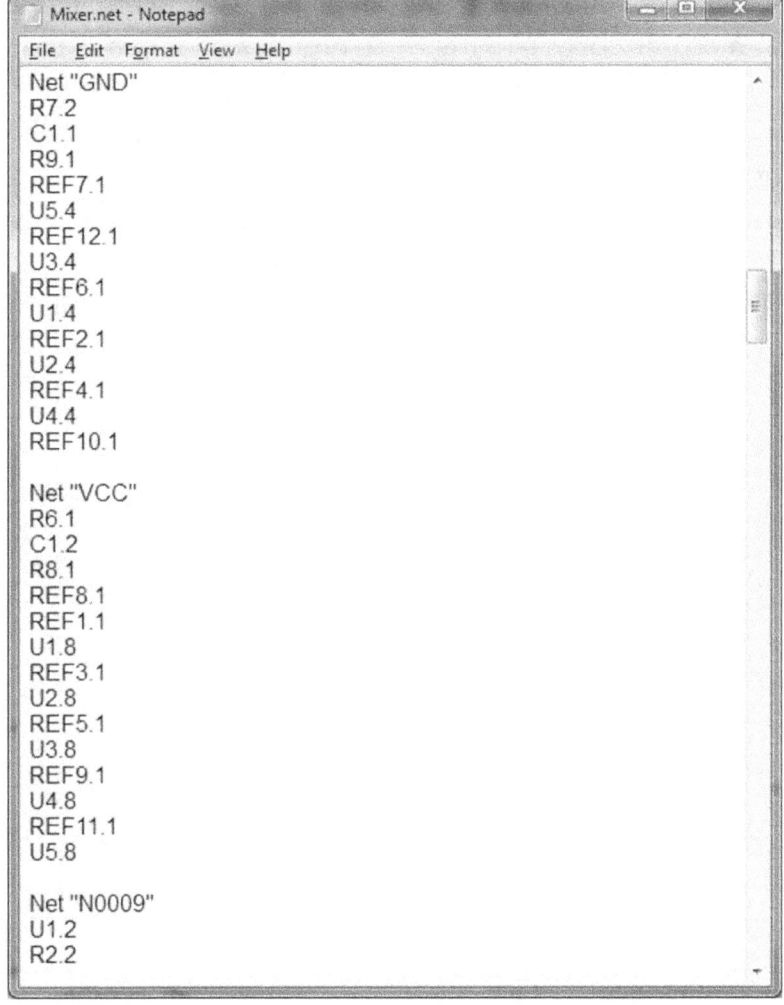

Net "GND"
R7.2
C1.1
R9.1
REF7.1
U5.4
REF12.1
U3.4
REF6.1
U1.4
REF2.1
U2.4
REF4.1
U4.4
REF10.1

Net "VCC"
R6.1
C1.2
R8.1
REF8.1
REF1.1
U1.8
REF3.1
U2.8
REF5.1
U3.8
REF9.1
U4.8
REF11.1
U5.8

Net "N0009"
U1.2
R2.2

We will skip Schematic ↔ consistency check for now. It is important when the PCB has been designed.

Unconnected Pins

This report can save your bacon, unless you are allergic to bacon, then it will save you from your bacon. The schematic is scanned and you will get a list of unconnected component pins. This is very common in designs. These unconnected pins can be very problematic. The obvious problem is that it may mean that your design will not work since the circuit is incomplete. Not so obvious is that in some circumstances such that of CMOS integrated circuits, if an input is left open the integrated may begin to oscillate due to feedback from the power supply back to the floating (isolated or unconnected) pin. This oscillation will cause circuit noise and in some cases cause internal heating and destroy the IC.

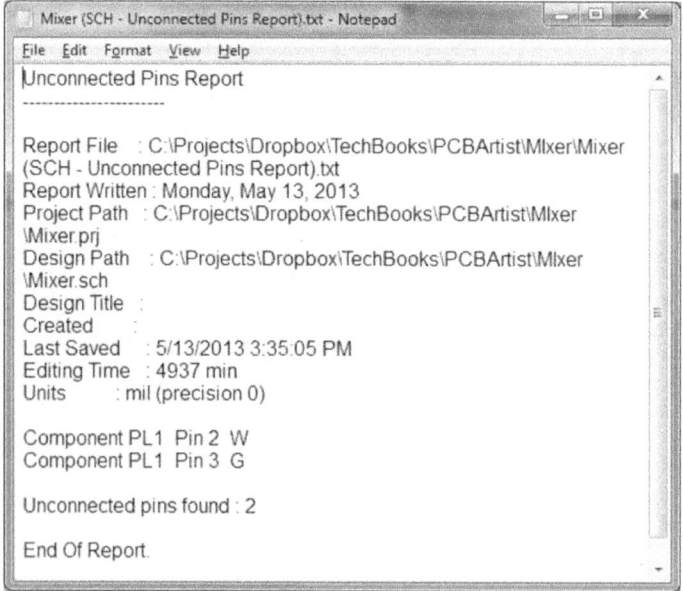

In our design we have opted to provide an option to power the circuit from a USB source. The two outer pins of connector PL1 are power (p1) and ground (p4). The two center pins have no function and are not connected. It should be noted that some devices do not power the USB ports unless there is a connection to pins 2 and 3. This scheme (design) will not work on those devices.

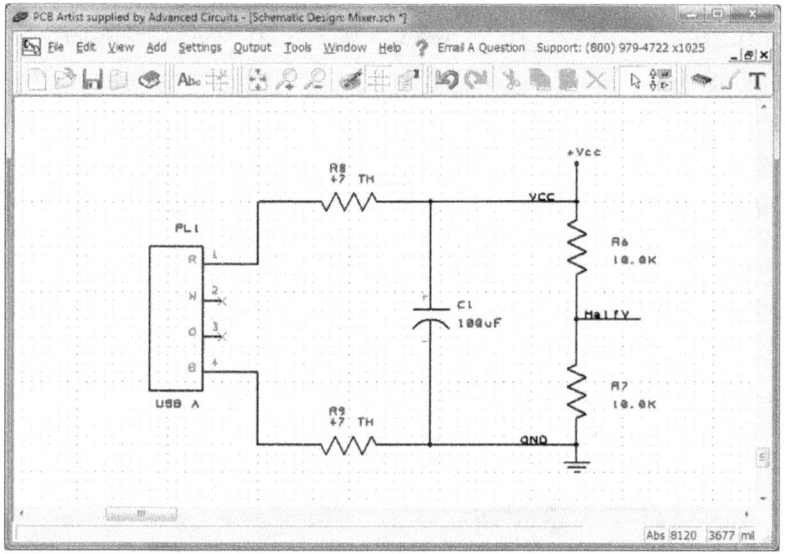

Should you feel Froggy (ready to jump) you can look up the resistors and how they connect to switch the USB output to high power. I do not want that capability on this board.

Entering the Schematic

Please keep in mind that excellent visual tutorials are on the web page and this is a supplement. It would be best to view the tutorials and then return to this section.

We now have the basic building block of the schematic and printed circuit board so we can begin the design. Type Cnt-N to open the New Design screen. Select the New Schematic Design button and use the pull-down menu to select default.stf. You can also select the Add to Open Project check box. This will open a file selection box and the schematic file will be created in the same directory as the project file.

Press the full screen button □on the Schematic window. Then input Alt-V A for View All. If you haven't added any items you will see the entire work space.

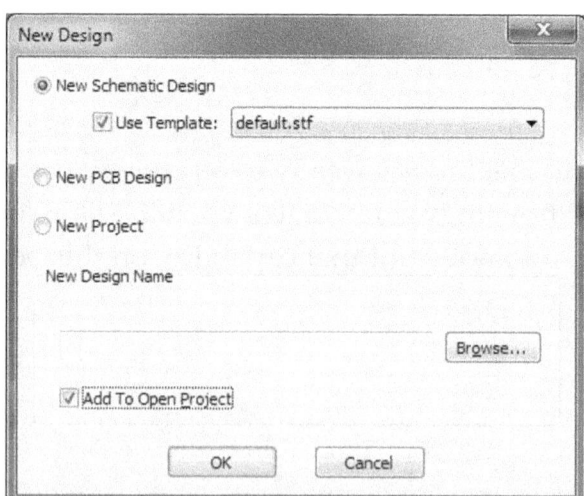

Placing a Component

Press F8 on the keyboard and you should get the Add component dialog box. Select the OPA2322AODR component in the Component box and press the Add button.

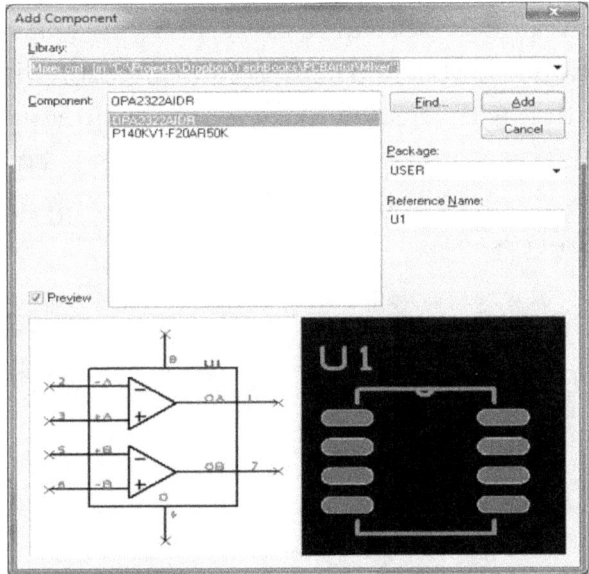

Assure that the Preview box is selected. You should notice that both the schematic symbol and PCB symbol are shown. Select the Add button.

This will return you to the schematic and you should have the schematic symbol attached to the end of your mouse cursor.

Position the component in the lower left corner of the schematic.

Left click once. The part should now be placed and not follow the cursor.

Press the escape on your keyboard.

Select the Cancel button.

Press Alt-V A and the display will zoom the active area to all the components in the schematic. Since there is only one component you should see a large copy of the op-amp we drew in the libraries section as in the illustration below. There are two op-amps in this package. Both are included in the symbol and the symbol is much like the physical package. However, if you look close at pins 5 and 6 you will notice that the schematic symbol is marked +B (non-inverting input of op-amp B) and that is connected to the – sign inside the triangle of the op-amp. This is an error! The pin number is correct, the Name (+B) is correct but the – and + inside the op-amp's triangle are reversed.

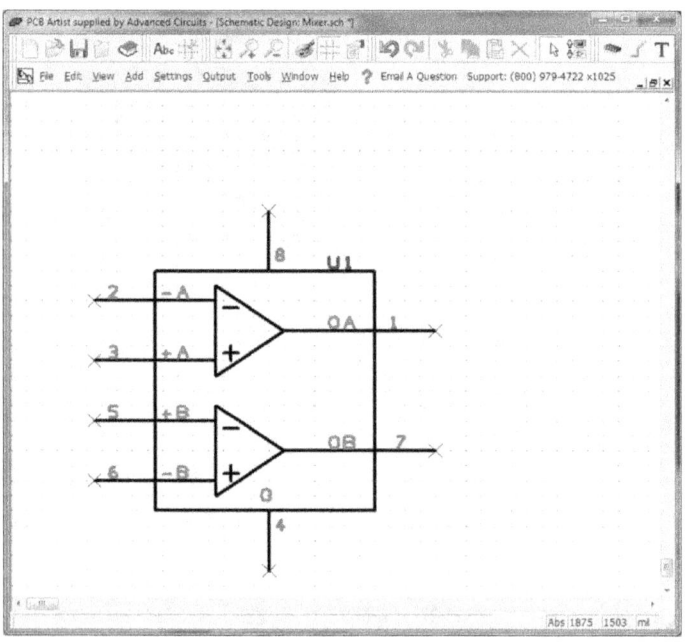

We only need to change the drawing inside the triangle.

Editing components in Schematic Editor

If there are changes to a schematic symbol it can be edited directly in Schematic Design. Right click on the symbol and select Edit Component in Library. If we examine our excerpt from the TI documentation to the right we note that pin 5 is the positive (non-inverting) output for the component. The label is correct and the drawing (the triangle and contents) are wrong.

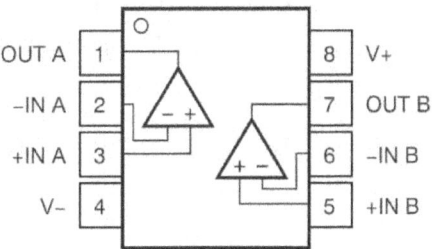

Select the component by clicking (once) on the component. Right click and select Edit Component In Library. You should get a screen much like the one below.

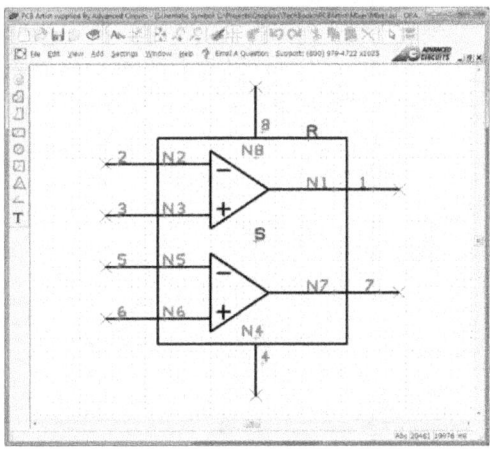

You can now click and hold on one of the lines that are used to draw the – inside the triangle. It will change color and you can move the – out of the way, but not touching any other line. Left click in an open area of the schematic and the line should deselect. This made room for the + to be moved down to where the – was positioned when we started.

Dual Selection

Then you can click on one line of the + and it will change color. Press Cnt and hold it while you press on the other line of the + and both the horizontal and vertical lines will change color, Move the + to where the – was originally.

Select (as above) the negative and put it where the – was.

It should look something like this:

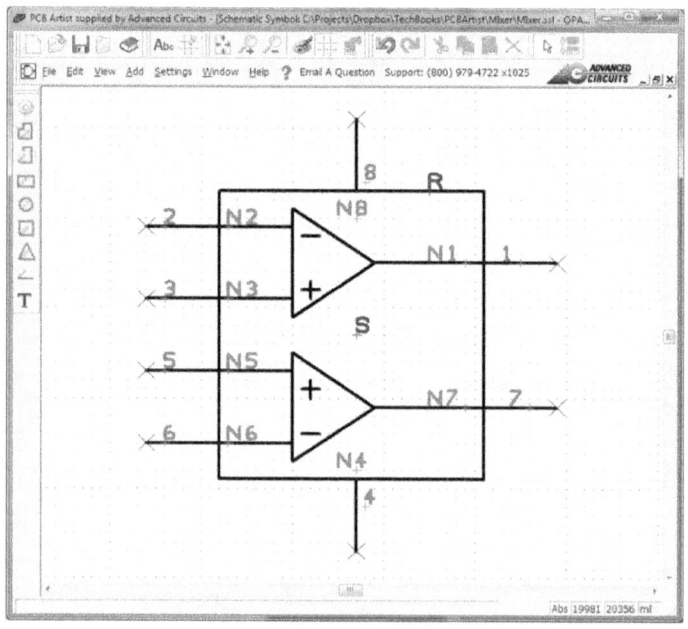

Notice that instead of +B and -B we have N5 and N6 respectively. This schematic symbol is general can be used with other components. The first step is to create the schematic symbol. The second step is to create the PCB symbol. Finally when both schematic and PCB symbols are prepared the component is created from the combination of schematic symbol (see the section on Creating Components) and PCB symbol. During the component creation process we instructed PCB Artist to substitute +B and -B for N5 and N6.

Type Cnt-S to save the symbol.

It will check that you want to save the component to the original name and library. Press OK. It will ask if you want to overwrite. Press Yes.

Type Alt-F S.

And we get this:

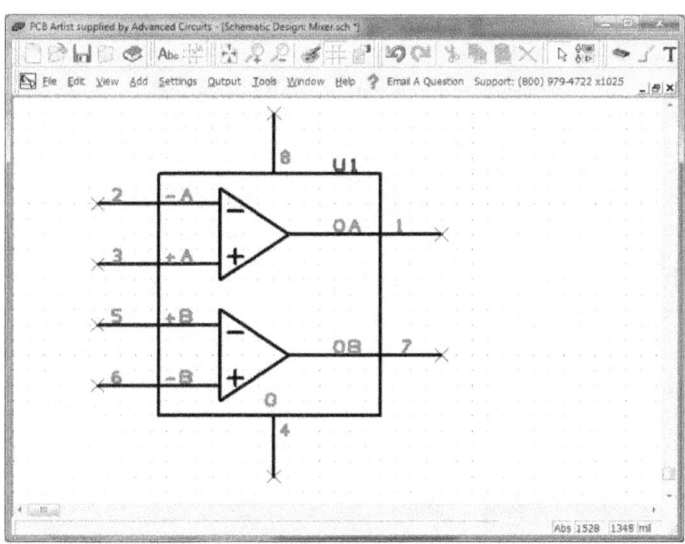

Wait just a minute! It's just like it was !!!@!!!!!#@#$%#$^^&&&!

That's because when PCB Artist originally gets the component from the library it makes a copy. That copy is put into local cache memory greatly speeding up the redraw process as well as assuring that changes in the library version of the component won't mess up our project. Even more importantly it assures that if we edit our local copy it won't mess up the original library.

Now we have to move the library copy into our cache. Type in Alt-T and select Update Components then select Browse giving the following screen. Select our part and select OK.

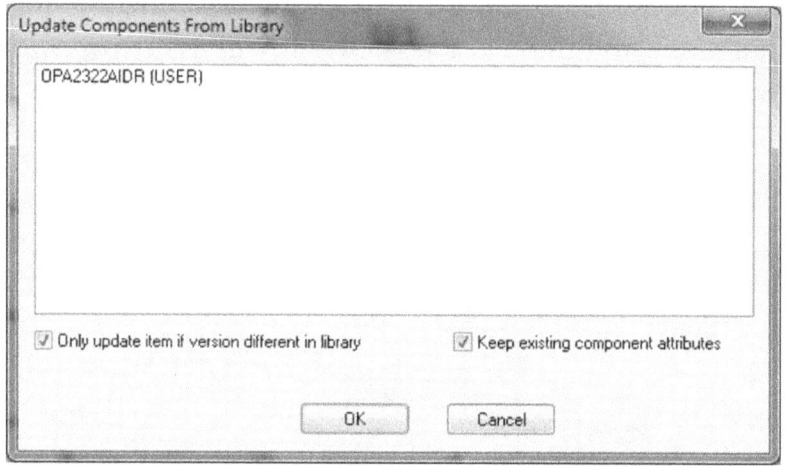

You should get the Update Components Summary screen.

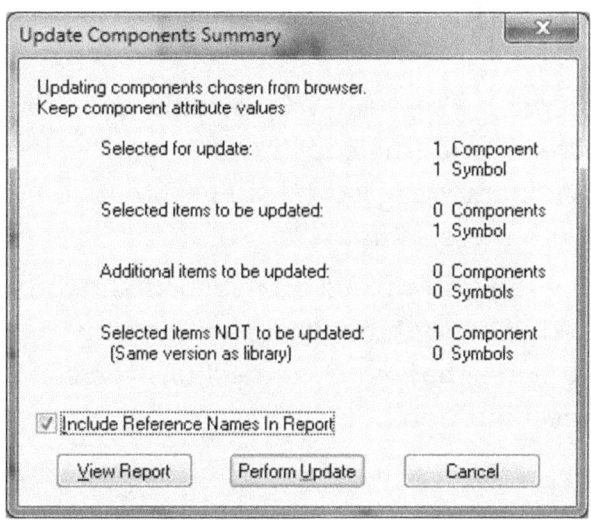

Select Perform Update.

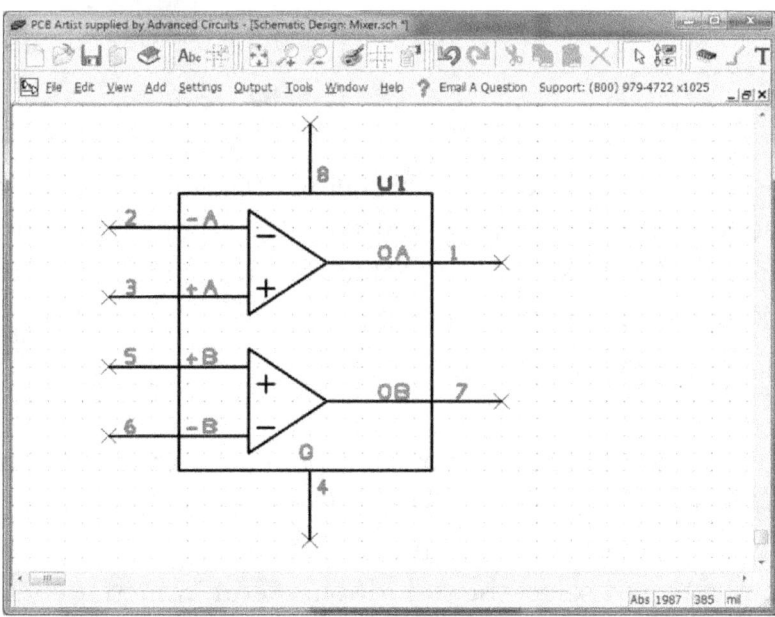

Ah, that's better!

Setting the Grid

This is essential. Press Alt-V O to zoom out . Press Alt-V and select Screen Grid (my copy doesn't work using Alt-V G as the pull-down menu suggests.) I suggest that you draw schematics with these settings.

If you leave the grid turned on in Schematic Capture you can be assured of everything lining up correctly. Otherwise you can have wires that won't connect or, worse, wires that seem to connect but don't.

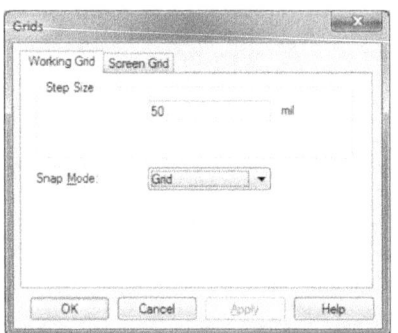 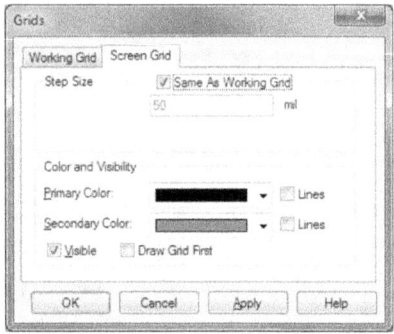

The problems associated with being off-grid are difficult to fix and sometimes result in re-entry of the schematic simply because it takes less time to reenter and check all those components than to fix grid problems.

Adding Components

Now we need to add a some components and wire them. Press F8 and select some resistors. When the Add component dialog box opens select the Find button. In the Find screen check Name and select the pull-down menu immediately to the right and select Is Exactly. Type in R and press the Find button.

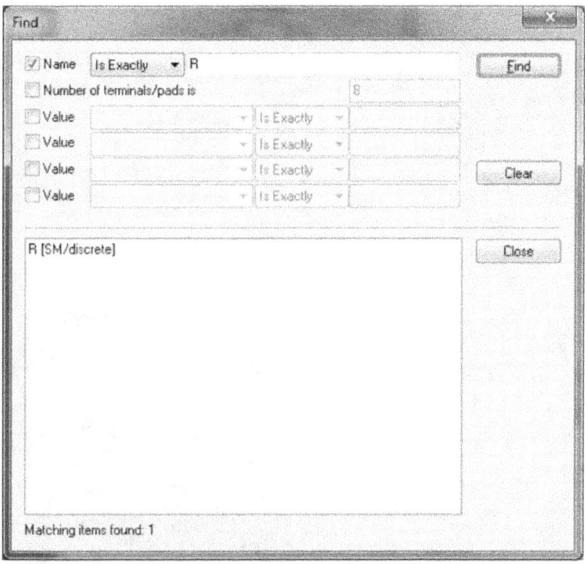

The result is R outside the brackets. Inside the brackets is the name of the library in which R resides. Select R (SM/discrete.) Select Close. You should see something similar to the following:

Select Add.

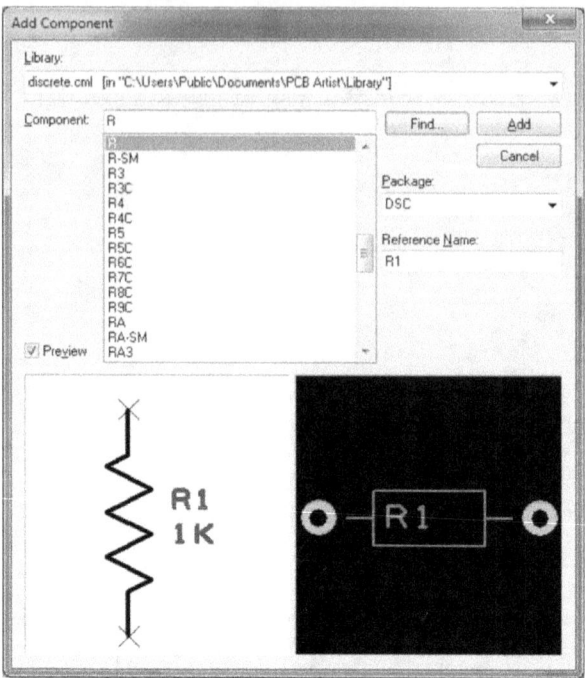

The symbol will be attached to the mouse tip and a copy will be added any time you perform a left click. After you have placed all the copies of R you wish press escape and you will return to the Add Component screen.

Search for and add the symbols Vcc and GND.

Using left click to select arrange the components in a manner something as follows:

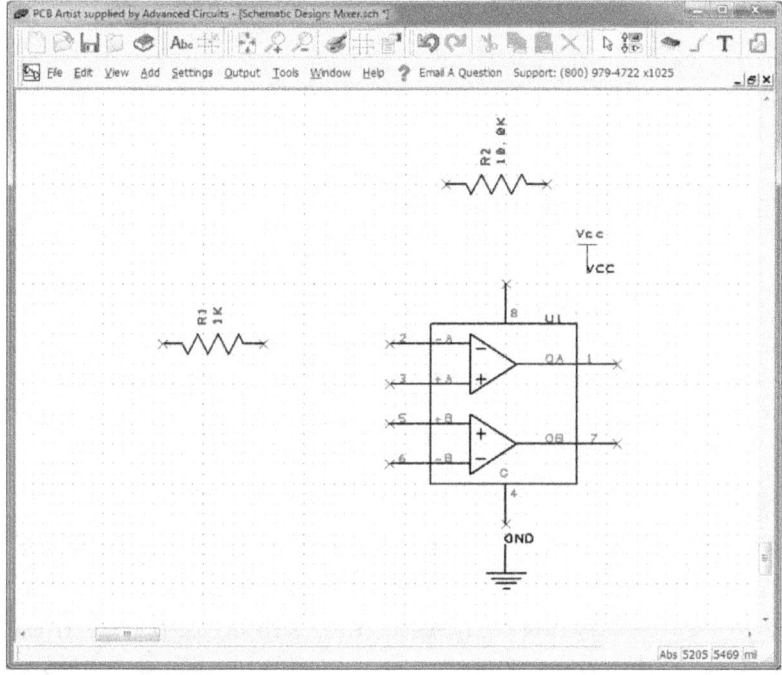

Wiring components

Notice that most of the pins of components have an x at the end of a line. Not all do and we have both types here. Connecting components consists of connecting the pins. One admirable feature of PCB Artist is that it strongly resists placing a connection (wire) unless it is connected at both ends.

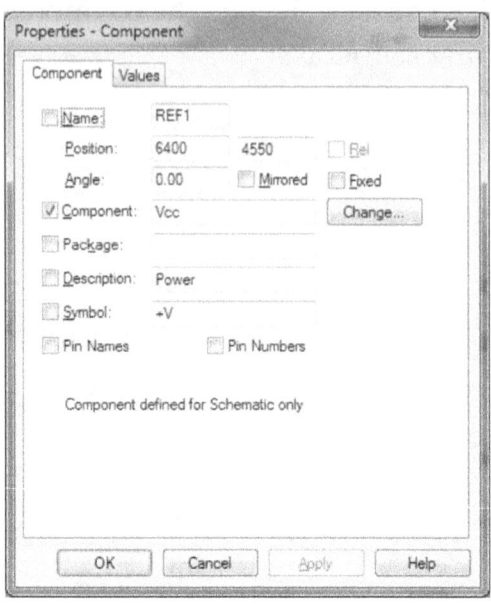

If you zoom in to the +5V symbol you will not be able to see the x and you will see that the symbol is not in the 'American' style. We can edit that component by selecting it (use the left hand click and assure that all the symbol changes color,) and selecting properties with a right hand click and selecting Properties near the bottom of the menu. Select the component tab and you should get a screen like the one here. Especially note that the Name is unchecked and component is checked. This is not a very versatile component.

To make this more to the US style we can edit the symbol shape to make it a ball instead of a line and use a Value instead of the Component name for its display.

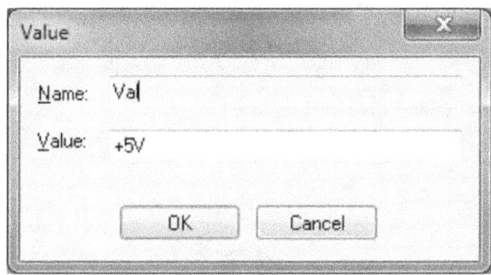

Uncheck component. Select the Values tab. Select the Add push button and type in Val for the Name: and +5V for the Value: as in the illustration. Select OK then select OK again.

Left click and select Edit Component in Library. In the lower left screen select the component and right click and select Edit Schematic Symbol.

Select the horizontal bar and press delete on the keyboard.

Click on the top end of the vertical line and pull it down making it short. This is a personal preference, but you can always make the effective line longer when wiring, but you can't make it shorter if it is in the symbol. With the line still selected right click and select Style. Select the Style to and set it to Normal. The style will then be set by the project. Deselect the line (click in a clear space.)

Now we can add a circle to the top shape. Type Alt → A and then C. Right Click in a clear space and deselect Define From Circle. Left click and hold on the top end of the line. Holding the mouse key down move the mouse cursor up to create a circle. Click to set the circle.

Select the R (reference) and press the keyboard delete key. Do the same for the N1 and the 1.

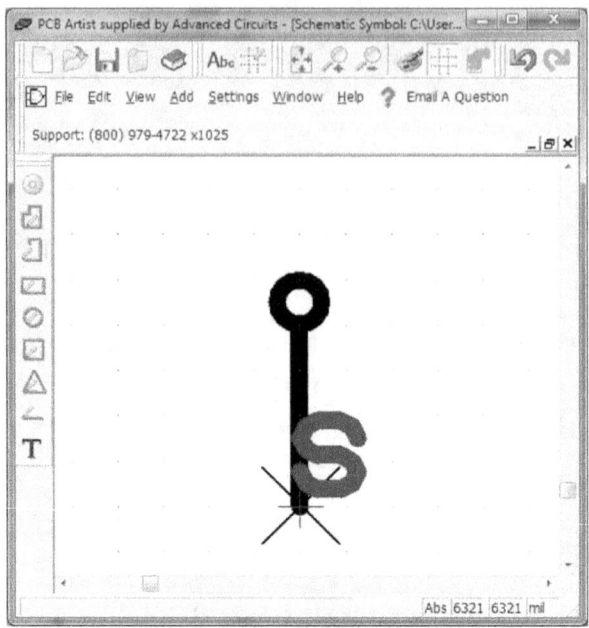

Save the schematic symbol and return to the Library Editor.

Finally, in the component editor type Cnt-E and select properties. Deselect the Reference check box under Names In Schematic. Since we want to use the Val property to designate the voltage source to which we are connecting, having the Reference also visible would muddy the waters.

One more thing. When you save the component call it Pwr and you can use it for any power source.

In the library editor the Net (Class) is defined for this pin. This would be okay except that many designs, especially microprocessors, have a number of power levels. This symbol might be connected to any of a number of points in the circuit. This might connect all your power voltages together. In this situation I'm showing the 'before' screen instead of the 'after' screen.

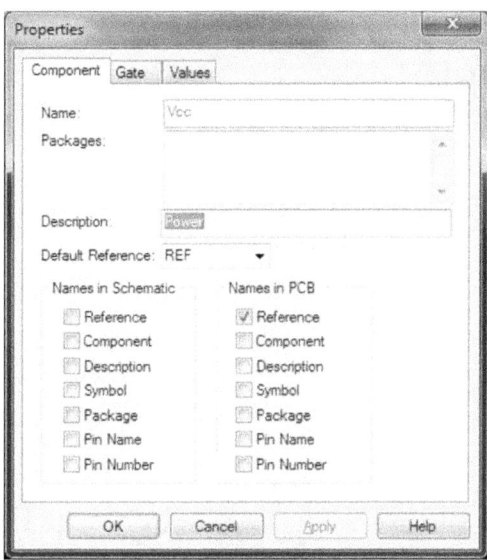

Save the library and return to the schematic drawing.

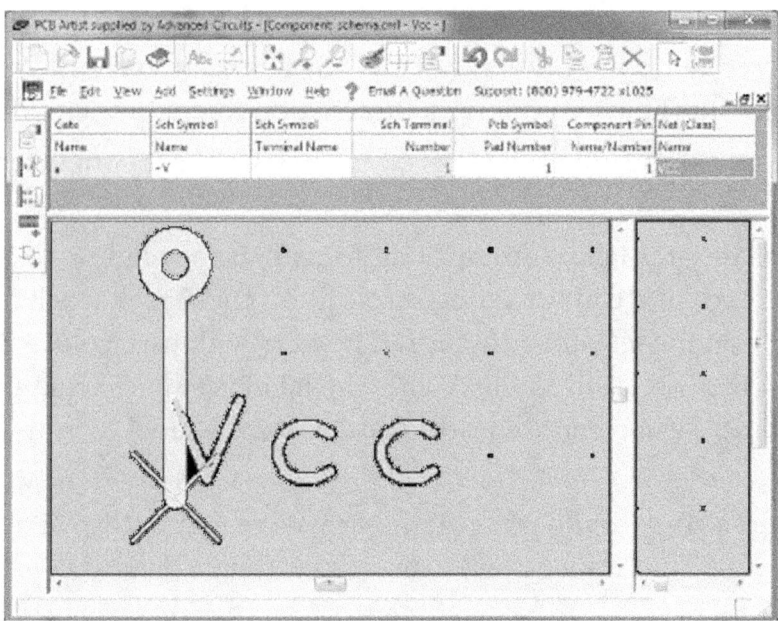

Your drawing should now look something similar to the following:

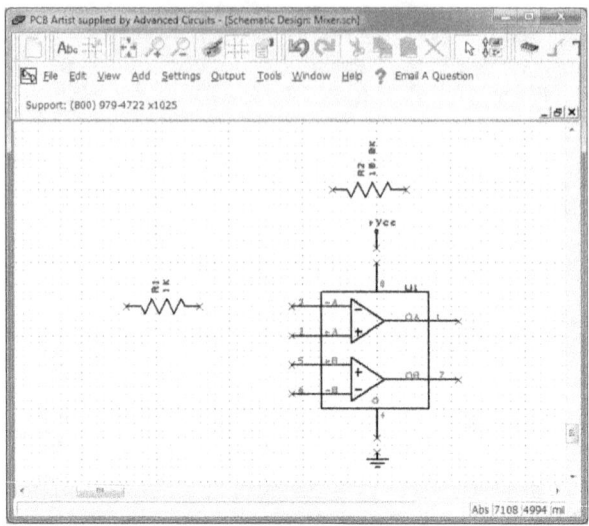

You can begin wiring (connecting) the schematic by pressing Cnt-A then O. Position the cursor over the x at the top of pin 8 of U1. Click once and drag the line to the X under the +Vcc symbol. Click again and the connection should be made with the wire no longer attached to the mouse cursor.

Press escape when the power and resistors are connected as follows.

The resistor labels are difficult to read at the angle presented. We can turn those around by clicking on each label, say R1 on the left most resistor. The label will turn color. Press R and the label will rotate. While the label is still selected press and hold the left mouse button and place the label as you wish.

Add a copy of the potentiometer we created in the Mixer library.

Please examine the schematic below and type Cnt-A then O and click the left mouse button to the left of pin 3 of the IC. This will begin a wire (connection) take that over to pin 3 and click on the x. Right click on the wire and select Display Net

Name. Double click on the name that just came up. It will be something like N0020. Select the Net tab and type HalfV into the Name: text box.

Place a wire begin pin 5 of U1 and the net HalfV.

This technique gives you a method to connect nodes (wires and traces that have a common metallic connection) without having a spaghetti collage of wires crossing and going everywhere. This is more important as the design becomes more complicated.

Your schematic should be very similar to the following schematic.

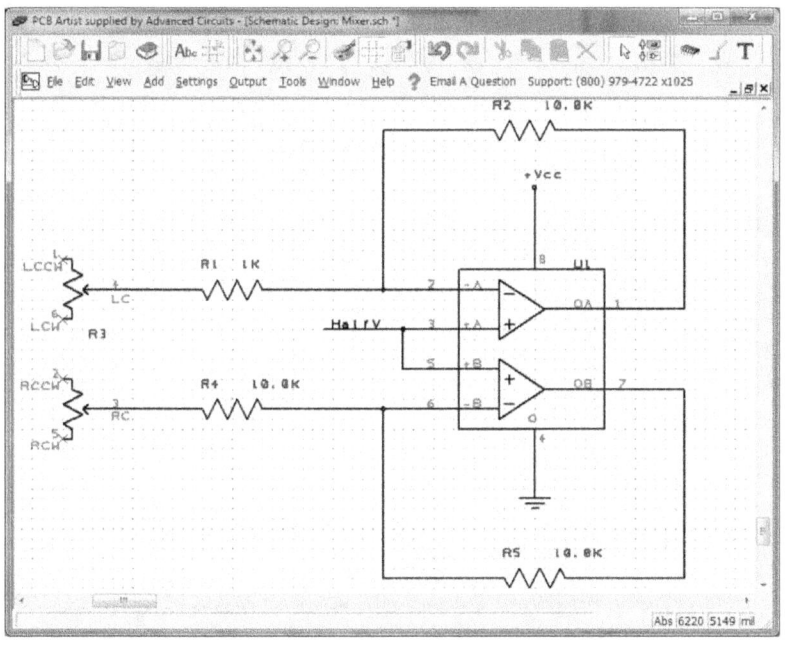

I have completed the schematic and attached it to this book. I have also placed the schematic and library on-line at www.AbrahamAaron.com/PCBArtist/Mixer.zip .

The full schematic view (Alt-V A) appears as in the Illustration that follows. The three outlines around components are sheets and can be found in the schema.cml library. I have included them in the Mixer.cml library. There are many unique things I like about this software and one of my favorites it the ability to create A size drawings from a large flat file such as this one. The attached schematic drawings are created by simply printing this file to a regular printer and selecting A (or letter) size sheet and 100% output. That is the method used to print the three attached sheets.

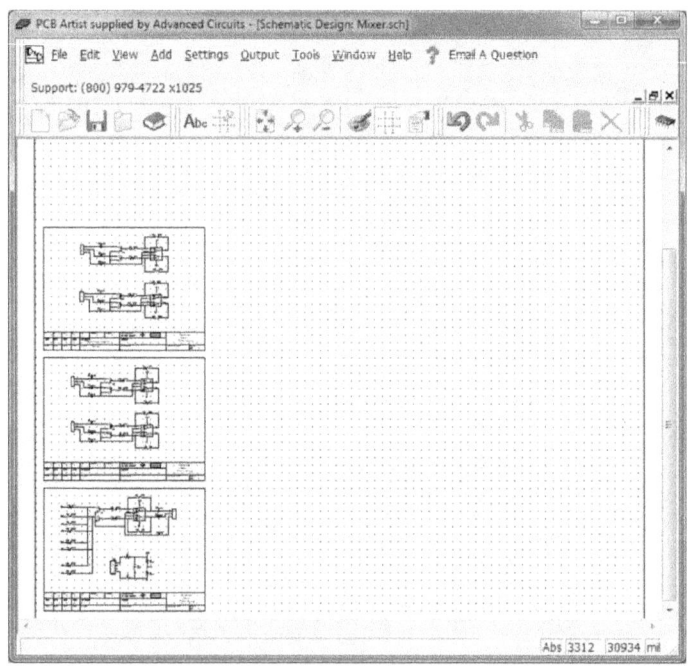

Bill of Materials

One of my favorite features of PCB Artist is their Bill Of
Materials reporting. In their truly infinite wisdom the authors
opened up the User Reports section to give us access to craft
and create our own reports.

Type Alt-O and R and select bill of materials from the screen
and then select Run from the right. This gives us the generic
bill of materials for the Mixer board. It is hard to read and
contains only minimal information.

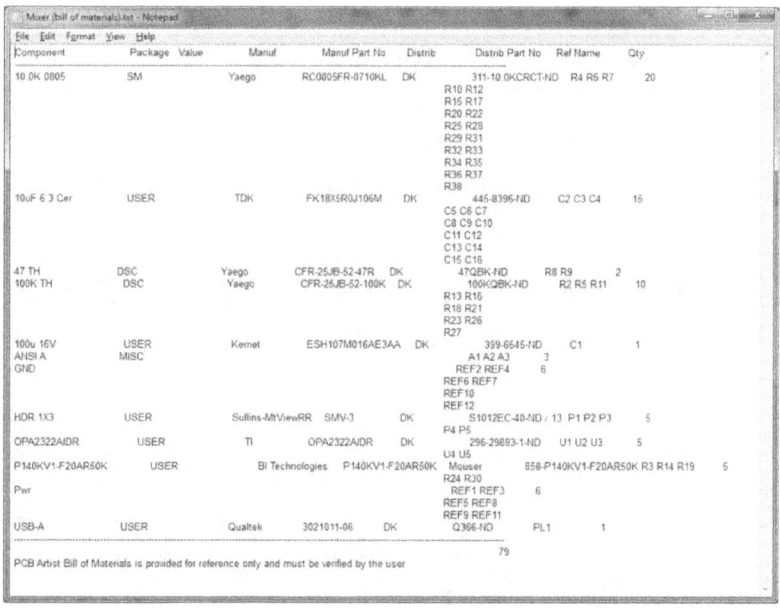

A more sophisticated bill of materials is included in the rear
of this book. The sophisticated bill of materials is also created
by PCB Artist. The report is put into a Comma Separated
Value (.csv) file and opened by a spreadsheet. In the past I
used Microsoft's Excel spreadsheet but became weary of

having the user interface change with every release. Therefore I use the Open Office suite. The price is right (whatever you want to pay) and the product is more stable than most.

http://www.openoffice.org/download/

Remember you can get a copy of the design files for this project from

http://www.AbrahamAaron.com

We entered some of that information as we developed the Mixer library. We will enter more of that info here.

Viewing Component Values

Please type Alt-S then V. You should get a list similar to the one on the next page. You may have to adjust the width of some columns. Note in the Illustration below that the mouse has changed to a doubled ended horizontal arrow indicating that the vertical line may be moved changing column width.

Name	Manufacturer	Manufacturer Part Nu...	Distributor	Distributor Part Number	De
A1					A s
A2					A s
A3					A s
C1	Kemet	ESH107M016AE3AA	DK	399-6545-ND	CA
C2	TDK	FK18X5R0J106M	DK	445-8396-ND	CA
C3	TDK	FK18X5R0J106M	DK	445-8396-ND	CA
C4	TDK	FK18X5R0J106M	DK	445-8396-ND	CA
C5	TDK	FK18X5R0J106M	DK	445-8396-ND	CA
C6	TDK	FK18X5R0J106M	DK	445-8396-ND	CA
C7	TDK	FK18X5R0J106M	DK	445-8396-ND	CA
C8	TDK	FK18X5R0J106M	DK	445-8396-ND	CA
C9	TDK	FK18X5R0J106M	DK	445-8396-ND	CA
C10	TDK	FK18X5R0J106M	DK	445-8396-ND	CA
C11	TDK	FK18X5R0J106M	DK	445-8396-ND	CA
C12	TDK	FK18X5R0J106M	DK	445-8396-ND	CA
C13	TDK	FK18X5R0J106M	DK	445-8396-ND	CA
C14	TDK	FK18X5R0J106M	DK	445-8396-ND	CA
C15	TDK	FK18X5R0J106M	DK	445-8396-ND	CA
C16	TDK	FK18X5R0J106M	DK	445-8396-ND	CA
P1	Sullins-MtVie...	SMV-3	DK	S1012EC-40-ND / 13	HD
P2	Sullins-MtVie...	SMV-3	DK	S1012EC-40-ND / 13	HD
P3	Sullins-MtVie...	SMV-3	DK	S1012EC-40-ND / 13	HD

Buttons: OK, Cancel, Add..., Edit..., Copy, Paste, Delete, Show, Hide, Revert, Colors...

NOTE: Information entered by means of the library can be transferred to other programs. Information entered by means of the Schematic (or PCB) editor is limited to the current project.

Looking through the screen you may notice that some components seem to be incomplete. Let's examine a few. A1, A2 and A3 are the symbols (schematic only) for the drawing outlines.

C1 is complete but C2 through C16 don't have entries in the Desc (Description) column. Double click in the the Desc column (or click and select Edit button from the right hand side of the screen) for C2 and type in CAP CER 10UF 6.3V 20% RADIAL. Click in an open (white) part of the screen and the value will be saved immediately. Select OK.

Name	Manufacturer	Manufacturer Part Nu...	Distributor	Distributor Part Number	Desc	Price	Val	Ripple	PartOf	Note	Taper
A1					A size Schematic sheet						
A2					A size Schematic sheet						
A3					A size Schematic sheet						
C1	Kemet	ESPU07M010AE3AA	DK	399-6545-ND	CAP ALUM 1000UF 16V 20% RADIAL	.19	100uF	135ma			
C2	TDK	FK18X5R0J106M	DK	445-9396-ND	CAP CER 10UF 6.3V 20% RADIAL	.48	10uF 6.3V				
C3	TDK	FK18X5R0J106M	DK	445-9396-ND	CAP CER 10UF 6.3V 20% RADIAL	.48	10uF 6.3V				
C4	TDK	FK18X5R0J106M	DK	445-9396-ND	CAP CER 10UF 6.3V 20% RADIAL	.48	10uF 6.3V				
C5	TDK	FK18X5R0J106M	DK	445-9396-ND	CAP CER 10UF 6.3V 20% RADIAL	.48	10uF 6.3V				
C6	TDK	FK18X5R0J106M	DK	445-9396-ND	CAP CER 10UF 6.3V 20% RADIAL	.48	10uF 6.3V				
C7	TDK	FK18X5R0J106M	DK	445-9396-ND	CAP CER 10UF 6.3V 20% RADIAL	.48	10uF 6.3V				
C8	TDK	FK18X5R0J106M	DK	445-9396-ND	CAP CER 10UF 6.3V 20% RADIAL	.48	10uF 6.3V				
C9	TDK	FK18X5R0J106M	DK	445-9396-ND	CAP CER 10UF 6.3V 20% RADIAL	.48	10uF 6.3V				
C10	TDK	FK18X5R0J106M	DK	445-9396-ND	CAP CER 10UF 6.3V 20% RADIAL	.48	10uF 6.3V				
C11	TDK	FK18X5R0J106M	DK	445-9396-ND	CAP CER 10UF 6.3V 20% RADIAL	.48	10uF 6.3V				
C12	TDK	FK18X5R0J106M	DK	445-9396-ND	CAP CER 10UF 6.3V 20% RADIAL	.48	10uF 6.3V				
C13	TDK	FK18X5R0J106M	DK	445-9396-ND	CAP CER 10UF 6.3V 20% RADIAL	.48	10uF 6.3V				
C14	TDK	FK18X5R0J106M	DK	445-9396-ND	CAP CER 10UF 6.3V 20% RADIAL	.48	10uF 6.3V				
C15	TDK	FK18X5R0J106M	DK	445-9396-ND	CAP CER 10UF 6.3V 20% RADIAL	.48	10uF 6.3V				
C16	TDK	FK18X5R0J106M	DK	445-9396-ND	CAP CER 10UF 6.3V 20% RADIAL	.48	10uF 6.3V				
P1	Sullins-MWire...	SMV-3	DK	S1012EC-40-ND/13	HDR 13 POS	.04			PRECOM05AA...	230 ht for 288 N use P...	
P2	Sullins-MWire...	SMV-3	DK	S1012EC-40-ND/13	HDR 13 POS	.04			PRECOM05AA...	230 ht for 288 N use P...	
P3	Sullins-MWire...	SMV-3	DK	S1012EC-40-ND/13	HDR 13 POS	.04			PRECOM05AA...	230 ht for 288 N use P...	
P4	Sullins-MWire...	SMV-3	DK	S1012EC-40-ND/13	HDR 13 POS	.04			PRECOM05AA...	230 ht for 288 N use P...	
P5	Sullins-MWire...	SMV-3	DK	S1012EC-40-ND/13	HDR 13 POS	.04			PRECOM05AA...	230 ht for 288 N use P...	
PL1	Qualtek	3021011-06	DK	Q366-ND	CB USB A-BLUNT CONN 6' 26/28 AWG	2.17	USB Ca...				
R2	Yageo	CFR-25JB-52-100K	DK	100KQBK-ND	RES 100KOHM 1/4W 5% CARBFILM T/R	.07	100K				
R3	BI Technologies	P140KV1-F20A450K	Mouser	858-P140KV1-F20A450K	POT ROTARY 10KOHMx1 14MM DUAL BUSH	1.08	50K				Audio
R4	Yageo	CFR-25JB-52-10K	DK	10KQBK-ND	RES 10K OHM 1/4W 5% CARBFILM T/R	.07	10K				

OK · Cancel · Add · Edit · Copy · Paste · Delete · Show · Hide · Reset · Colm...

Component Values

You can personalize the Report by editing at the Component symbol editor level. An important item for a potentiometer is referred to as its taper. For a linear potentiometer the resistance between the center tap and one end of the potentiometer will change approximately half the total resistance if you turn it to mid position or 25% of the total resistance if you turn it 25% of travel.

This type potentiometer is not preferred for audio work because our ear's response to sound is non linear. For low volumes doubling the sound power only increases the apparent sound level (what we hear) a small amount. In a circuit this means the first small increment in movement makes the sound much louder. An audio taper potentiometer may be used to correct this effect and improve the design. So let's add a field called taper to the potentiometer in our library.

Type Alt-F L and select the Components tab. Select the library Mixer.cml. Select the component P140KV..... in the Library components section and select the Edit button. Press Alt->Enter and select the Values tab. You should get the following screen.

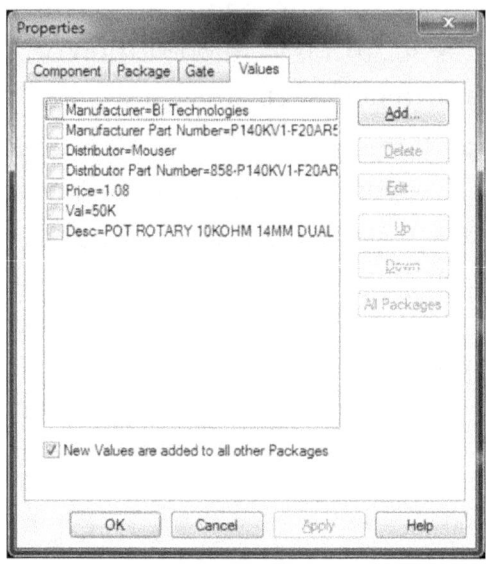

Select the Add button and type Taper in the Name: field and Value: Audio in the Value field. Select the Apply button and you should get the following screen. Now you have the information in the schematic symbol.

Select OK. Save the library while exiting.

Return to the schematic and type Alt-T and select the Update Components and select all components and select the OK button in the resulting screen. Press Cnt-S to save the schematic.

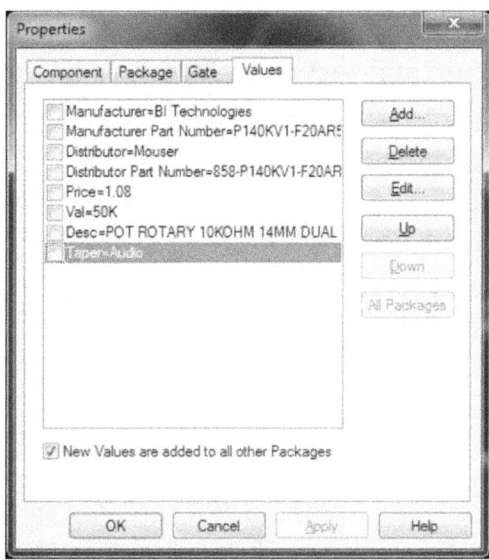

Type Alt-O then R and select Mixer and select Edit on the side. You should have the screen below.

Component List:

The Component List Report is divided into sections. The General section (undesignated) controls the report. The Components: section interesting, but I can't determine it's utility. The Columns in the Report determines the format of the report. The Sorting: section determines the order in which information will be is displayed.

You may notice that I've added a Mates With Column. Once you've researched a component you don't want to have to start over when you need to get the mate, say an obscure connector. So I put a reference (usually a part number) so that I don't have to start the search over again completely.

Select Edit and you should get this screen. From the top assure that CSV Format and Include Column Captions are checked. Select Add and Value from the Field: pull-down list.

This can be very confusing because with this software Value and/or value can have many connotations. In this case we want to output the (text) value of the Values: section at the bottom of this screen. Remember, I'm just the messenger.

We want to report the Taper of a potentiometer so type Taper into Caption. Logarithmic can be abbreviated to Log and Linear can be abbreviated to Lin so we only need to have a Width: of 5.

You can now select the Add button beside the Values: section of the Column Settings screen and select taper.

Another use might be to put your proprietary part number on your company's products.

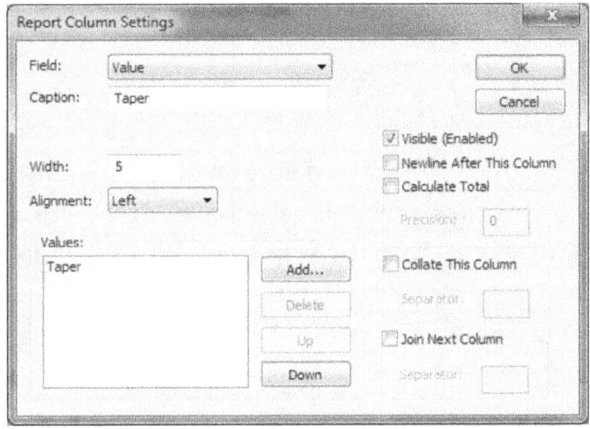

This is the technique you can use it if you dare. The rewards are great when it is close to the end of a project and there is a time crunch and you want to update the part number for 30 components.

Editing the PCB – Placing components

When we created the project in the chapter Create A Project. Now, the models are finished and the schematic is finished and the requirements are finished. Right? Let's edit the PCB.

Type Alt-T then F. You will get the Forward Design Changes screen.

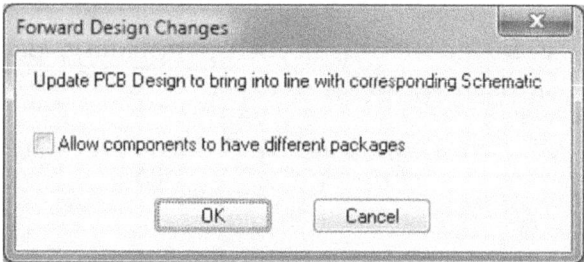

Select OK . You will get two screens. One is a text screen informing the changes made by the software. This is only useful as a debugging tool, but it can be invaluable.

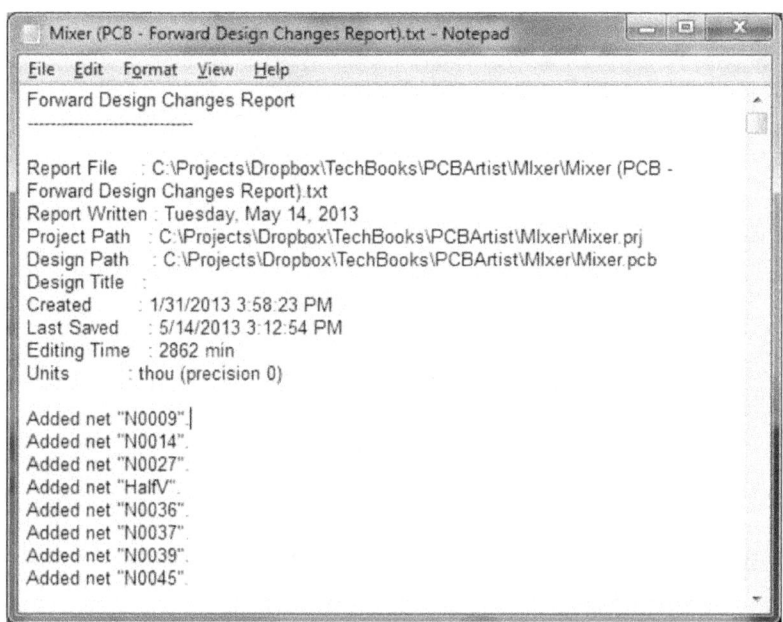

You also get a board layout.

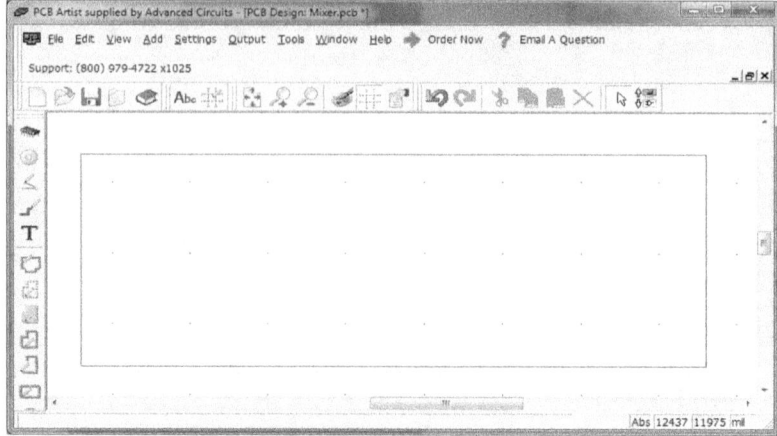

Where are the components?

Type Alt-V then A and you will see them bunched up down in the corner of the screen.

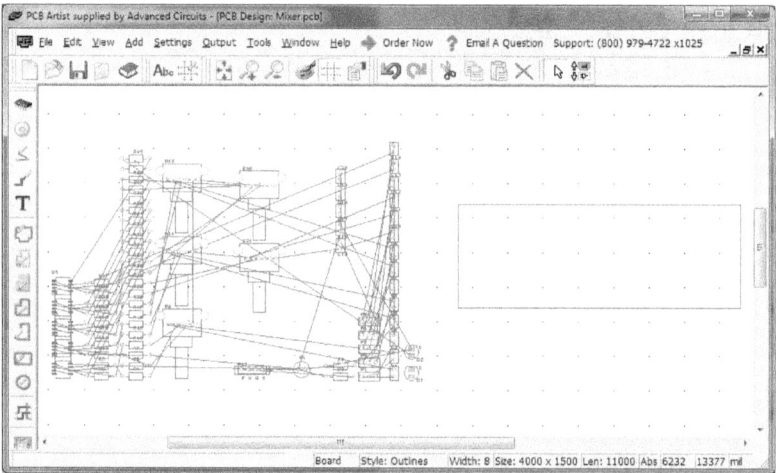

This is very important. Type Alt-S then G and set the grids.

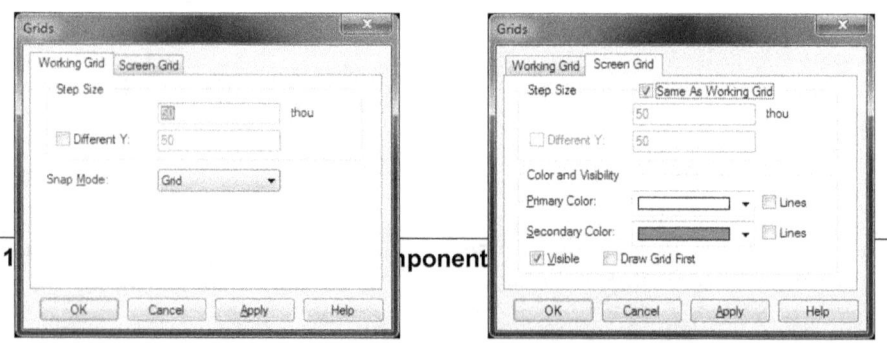

The difference above is that the left side has the Working Grid tab selected and the right side has the Screen Grid tab selected. This places the components on a fixed and relatively open grid.

I like to place the difficult components first for a sanity check. In the case of this board I believe the difficult components will be the potentiometers because of their size. When placed near the board it looked like the following.

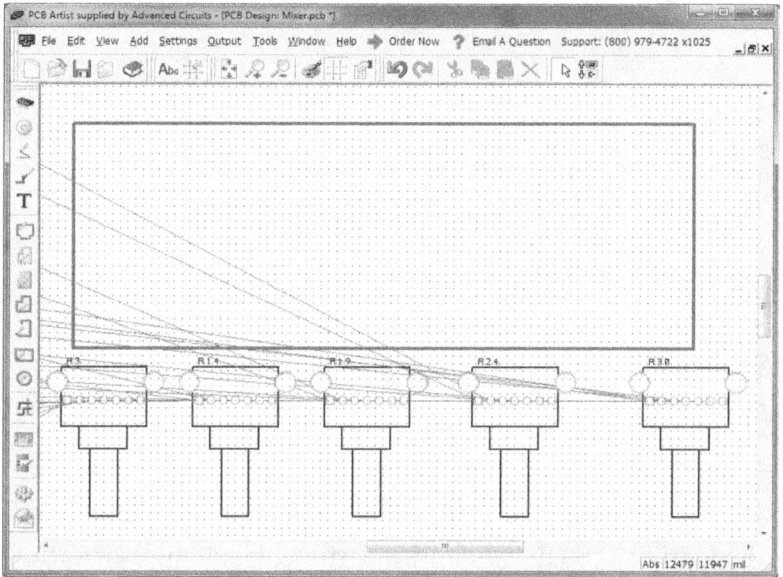

You should take note that when lined up, the potentiometers are wider than the PC Board. Houston, we have a problem! But it is a small one. Select the right side of the board and pull it to the right. Then I selected the left side and pulled it left. I have clicked on one of the lines of the PCB and it is selected (turned white) selecting the entire box. I would like to point out a number of items.

Notice that there is a panel at the bottom of the screen that is denoted Size: giving us the size of the PCB 5000 X 1500. The PCB is now 5" by 1.5". We can leave it there for now. Also notice the four panels to the right. Abs 12585 11640 and thou. The thou indicates we are working with thousandths of an inch (0.001".) The current cursor position (not shown) is at x coordinate 12.585 and the y coordinate is at 11.640".

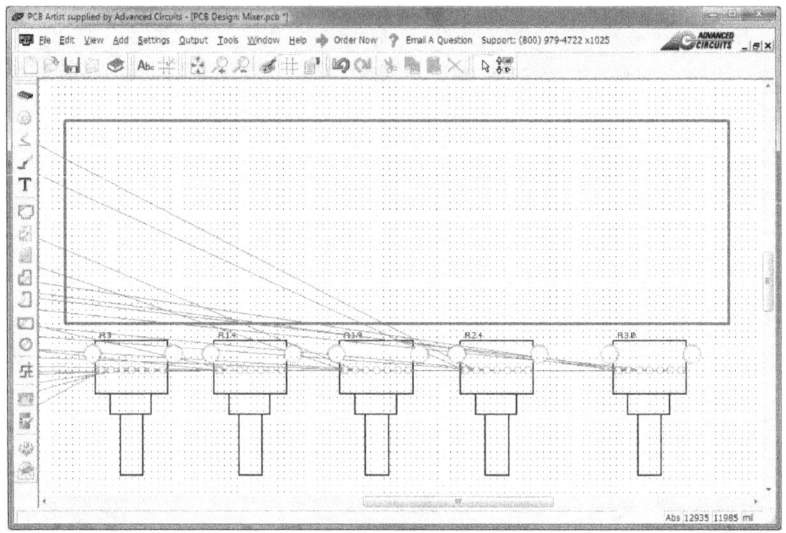

It would be convenient to have the lower left corner of the PCB at location 0,0 (this is and assumed Cartesian notation with the x (left and right) position first and the y (up and down) second. When the drawing begins the origin is in the lower left corner of the workspace. This allows the software designers to not concern themselves with negative x or negative y values. Convenient for them, but not for us.

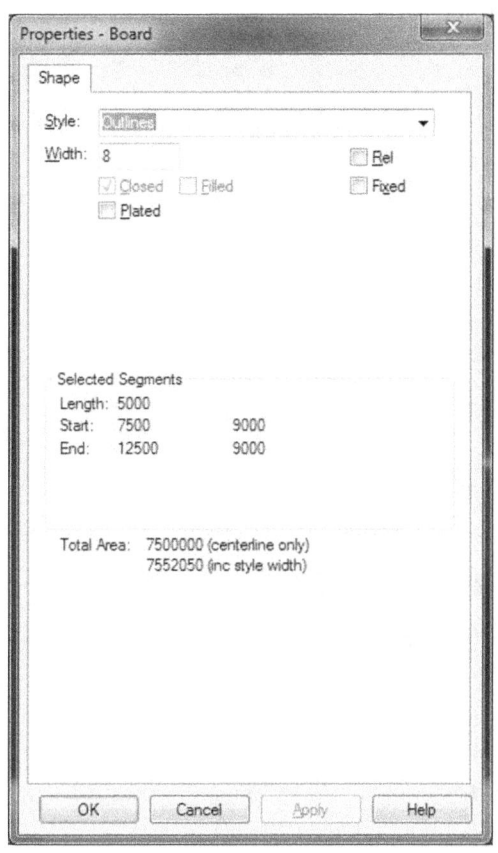

We can set the relative coordinates to be at the lower left corner of the board so that we don't have do a lot of math in our heads (sure I do it in my head.)   Left click on the outline and select properties. The board outline is drawn as a rectangle with the lower left and upper right corners defined. In the screen to the right we see my board values. Your board might have different values. We note the Start: line under selected segments. This is 7500 and 9000 (again implied x and y coordinates.

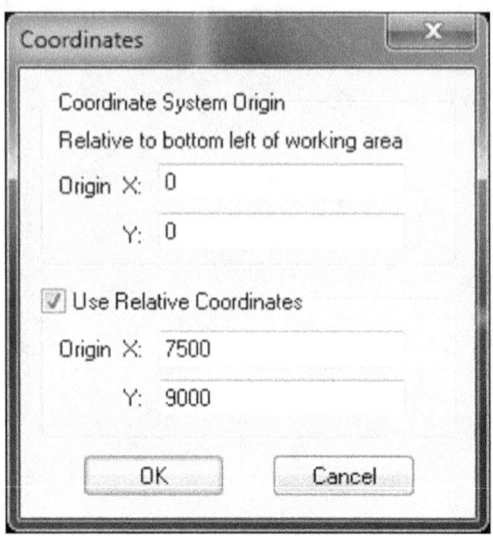

Cancel the Properties – Board screen and type Alt-S then C. Type 7500 in the Origin X: and 9000 in the Origin Y: box. Select OK.

Zoom into to the lower left corner of the board and you will see a cross in a circle (this is a traditional fiducial mark) exactly on the edge of the board.

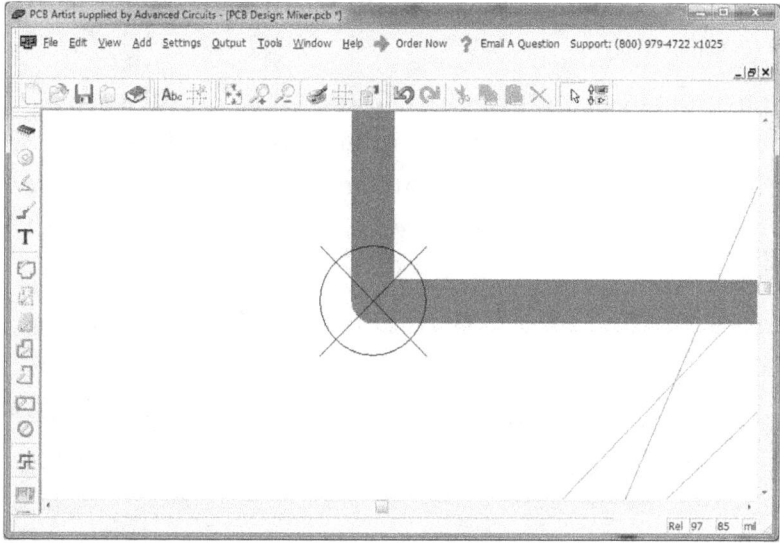

This is a good time to do a file save.

Type Alt-V then A to see the entire schematic.

Once upon a time, when HP was the bastion of technical correctness, (no sarcasm, HP was at one time the most accurate instrument company in the world) there was an HP document that suggested that schematics and, where possible, instruments should be arranged left to right and front to rear. I was working in the Nuclear industry and we took that document to heart. Today's HP is a truly different story.

I try to make my schematics read left to right – input to output.

Referring to the schematic. Please get the most recent copy from the www.AbrahamArron.com site. There are four input pots R3, R14, R19 and R24. These will be assigned channels 1, 2, 3 and 4 respectively. R38 will be the output level pot and place to the right.

An important function of the modern PCB software is an autorouter. The magical autorouter will do all the work of placing routes on the PCB. However, we can help the autorouter wire the board by consistent placement of certain components.

Press Alt-T on the keyboard and select Forward Design Changes.

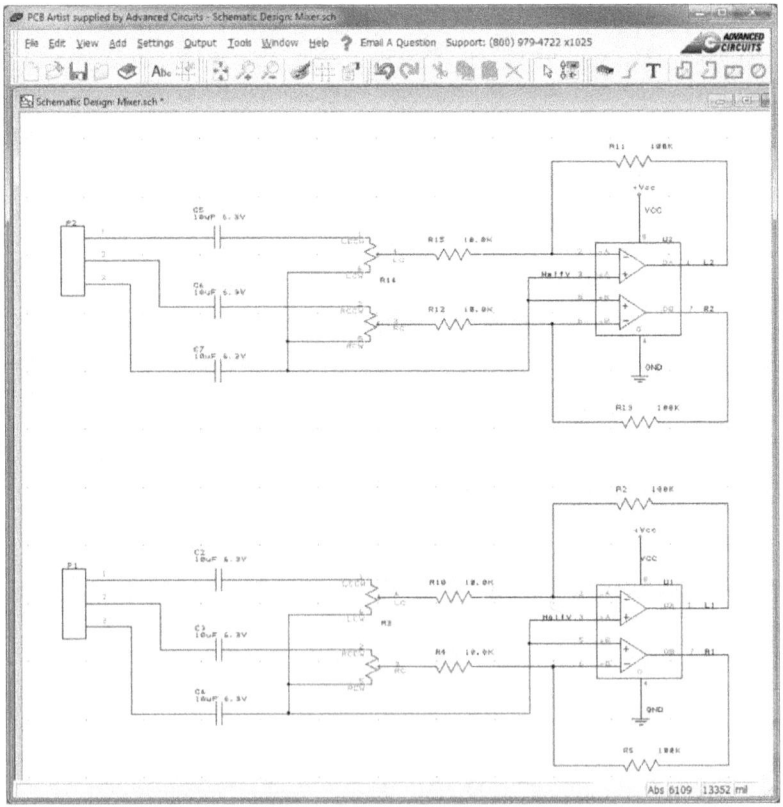

We will start by placing and routing the schematic section above (included for reference). The full schematic is included at the end of the book.

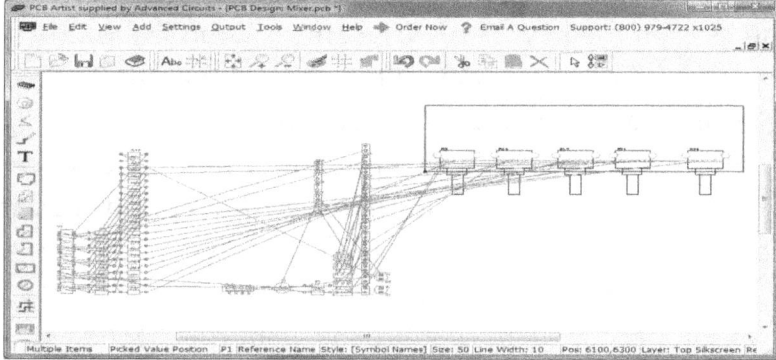

Sometimes the components can be difficult to find. Press F9 and you will get the interaction bar. From the top select Component from the first pull-down menu. Then select Goto from the bottom tabs of the Interaction Bar. You should get a display like the one below.

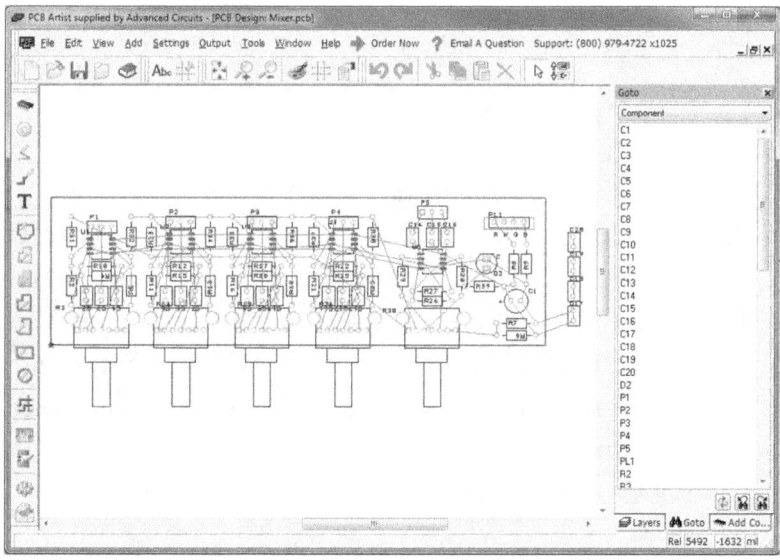

Channel 1 consists of P1, C2, C3, C4, R2, R4, R5, R10 and, of course, U1. Begin by selecting each in turn and placing them above potentiometer R3. C1, C2 and C3 are relatively obvious as they have one pin connecting to R3. Select each one in turn and press R on the keyboard. The entire symbol should rotate. After placing them near R3 you may notice that the reference designators are upside down. Select only the reference designator by clicking on it and press R until the reference designator is right side up and place the reference designator on top of the symbol.

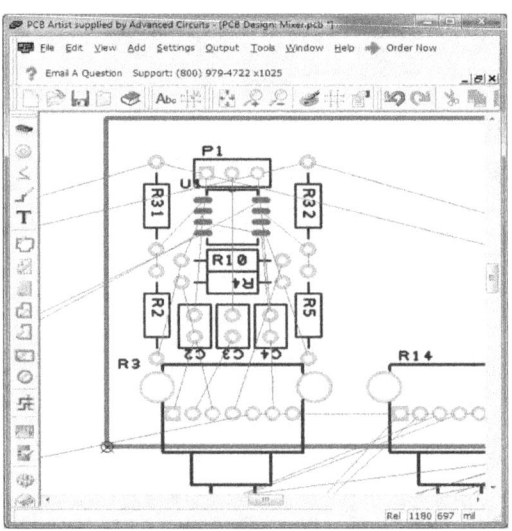

The traces from the bottoms of C2, 3 and 4 will go directly to the pot but the traces from P1 to the capacitors must go through that maze of parts. Yet I scrunched them down to the bottom of the board. This is because I will have to move the audio outputs across the board and I will have to distribute power and ground the length of the board. I've chosen to leave room at the top of the board for that purpose.

Also note that the IC is surface mount. My placement will allow unimpeded horizontal routing under the IC. More about power and ground later.

Lay out the remaining three input channels in a similar manner.

After you have placed all the components inside the board and before you straighten things out type Alt-T and select Design Rule Check (your greatest friend.)

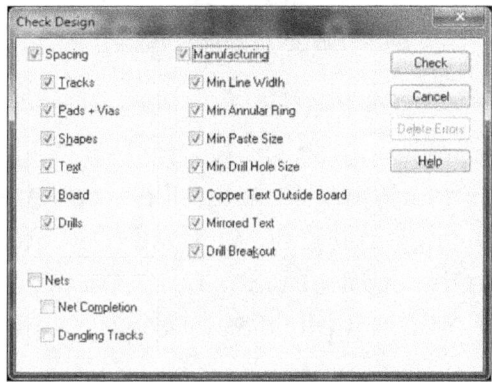

Check the boxes marked Spacing and Manufacturing each of which will mark all the check boxes below them. No need to check nets, there are none connected and you will get tons of errors.

Select the Check button. Hopefully, you will get no errors.

Error Correction

I just happened to get an error. Imagine that. There are two indications of error, one is a written text file that is opened automatically and the other is an indication on the printed circuit board's layout screen.

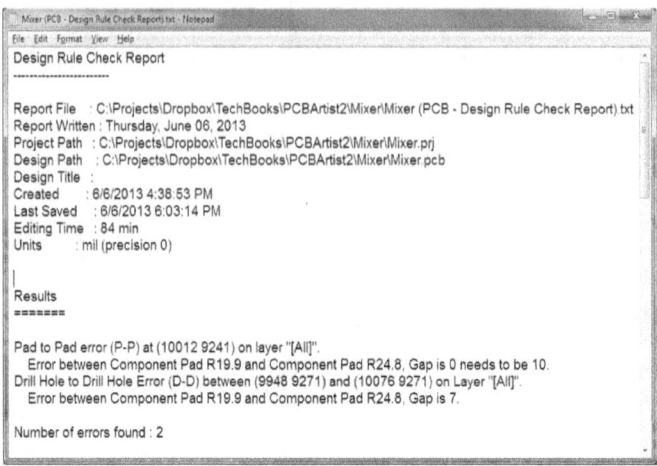

Design Rule Check Report

Report File   : C:\Projects\Dropbox\TechBooks\PCBArtist2\Mixer\Mixer (PCB - Design Rule Check Report).txt
Report Written : Thursday, June 06, 2013
Project Path   : C:\Projects\Dropbox\TechBooks\PCBArtist2\Mixer\Mixer.prj
Design Path   : C:\Projects\Dropbox\TechBooks\PCBArtist2\Mixer\Mixer.pcb
Design Title   :
Created   : 6/6/2013 4:38:53 PM
Last Saved   : 6/6/2013 6:03:14 PM
Editing Time   : 84 min
Units   : mil (precision 0)

Results
=======

Pad to Pad error (P-P) at (10012 9241) on layer "[All]".
    Error between Component Pad R19.9 and Component Pad R24.8, Gap is 0 needs to be 10.
Drill Hole to Drill Hole Error (D-D) between (9948 9271) and (10076 9271) on Layer "[All]".
    Error between Component Pad R19.9 and Component Pad R24.8, Gap is 7.

Number of errors found : 2

The screen representation shows the P-B (Pad to Board) indication and a nearly invisible line between the pad and the board. This indicates that there is a pad to board spacing error. I've made the error obvious but sometimes it is difficult to find the error on a complicated board.

This is where the error file becomes helpful. There is a header in the file that gives the time, date, project name....... and there is a Results section. The results section indicates the actual errors. The settings section indicates which of the Design Rule Check (DRC) options were enabled for this particular file result.

The errors in the Results file give the error type; the entity (line, pad, via, board....) location; layer name; English statement of the error and specific error parameters. I have always been able to find the error with this information when I evaluated this information correctly.

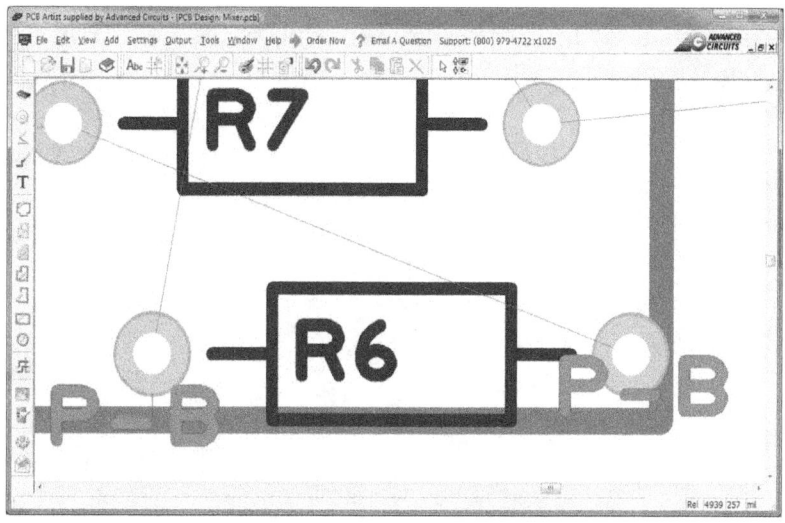

The solution is to select the component, remembering to select the entire component as indicated by its color change, and move it back into place. It is easy to select just a pin or just the reference designator.

Run the check again and the board should have a clean bill of health. Note that deleting the text reports between error checks will keep your screen clean and you won't end up chasing a problem that you've already fixed.

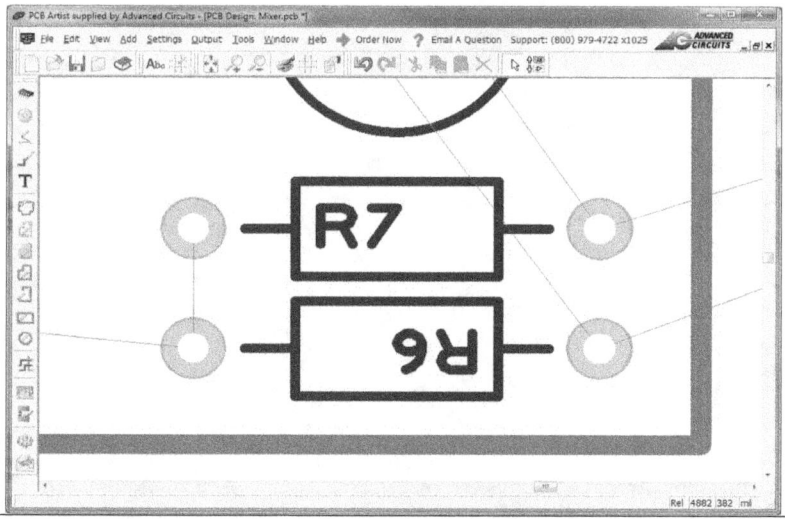

It's always a good idea to save the file before you run a Design Rule Check.

Editing the PCB – Routing Components

We first have to perform some housekeeping. Type Alt-S and select Preferences. The following screens are displayed to assure we are starting from the same place. These should be the default settings though we may change them in the future. I suggest that you check them all.

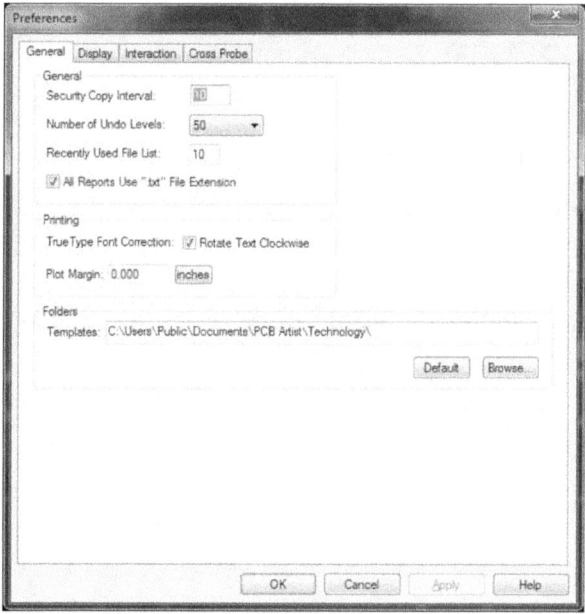

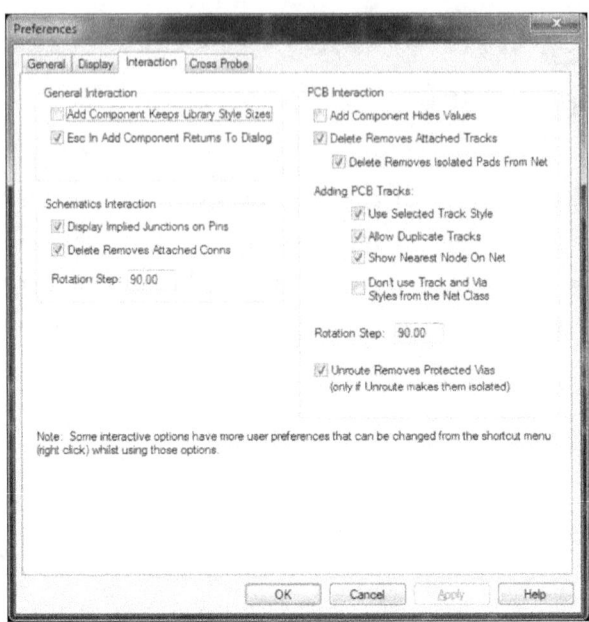

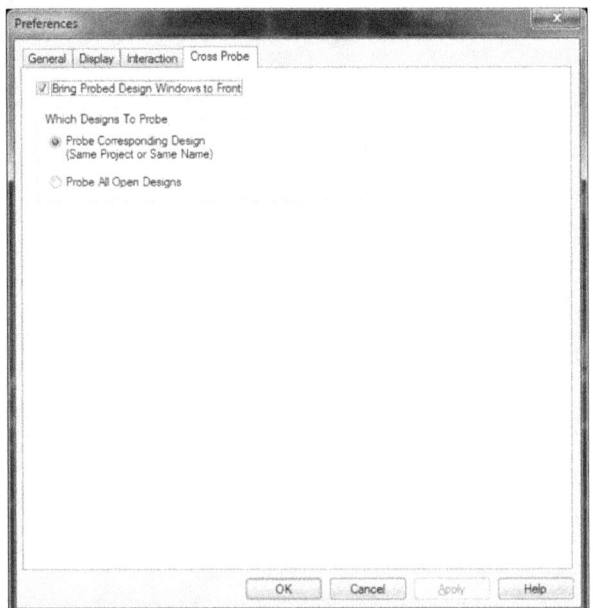

When we are changing component location the 50 thou grid is a good starting place for through hole components. However, for placing tracks a smaller grid is appropriate. Type Alt-S then G and use these settings for this tutorial.

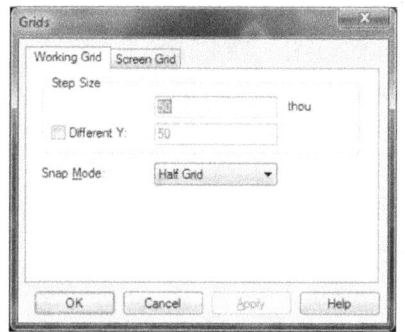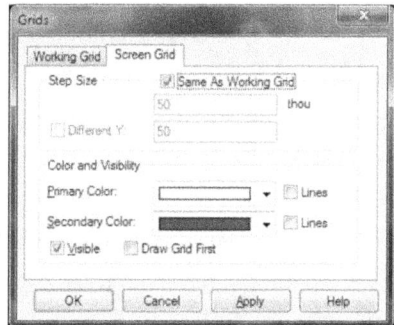

Type Alt-S and U and set the Units to thou with a precision of 0.

We recently set the system Origin but let's check it, just in case. Type Alt-S and select Coords.

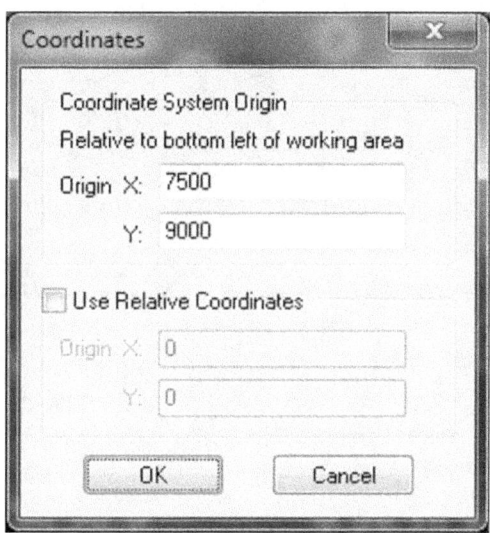

Type Alt-S then S. This section of the Settings is EXTREMELY important. In each case select the tab and select Delete Unused. Set the values as indicated in the following screen shots.

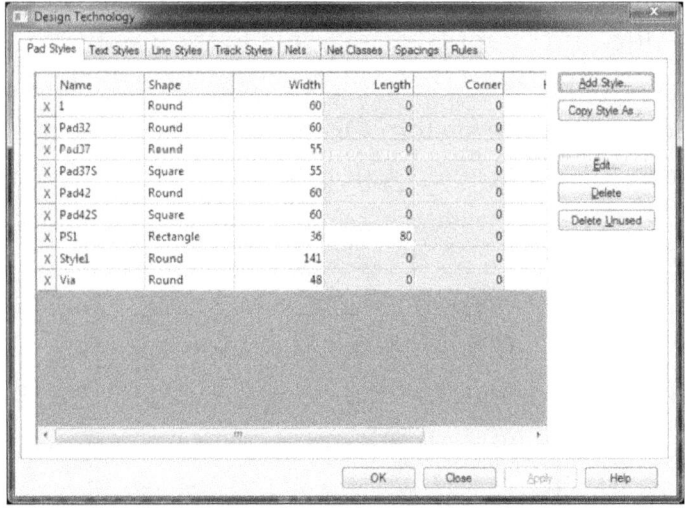

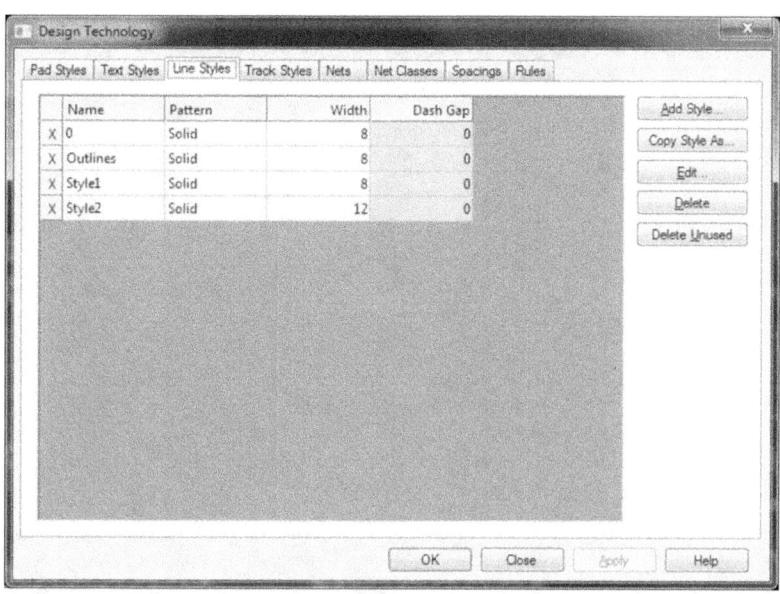

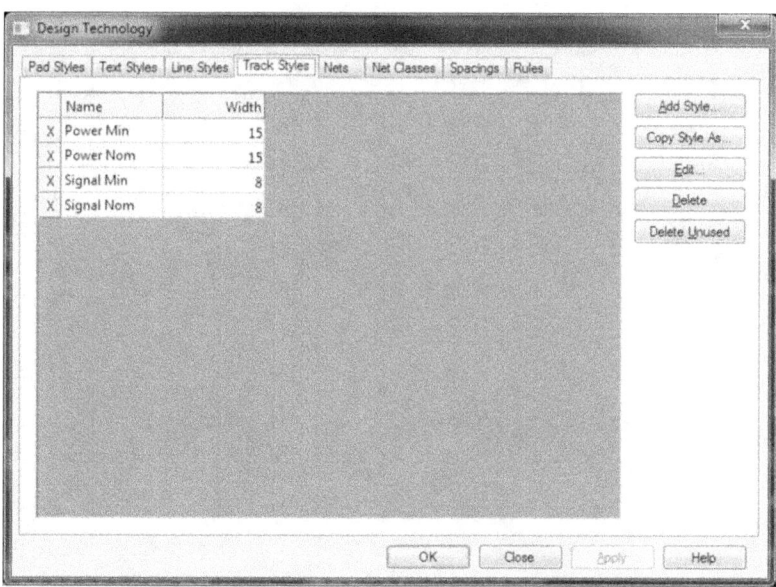

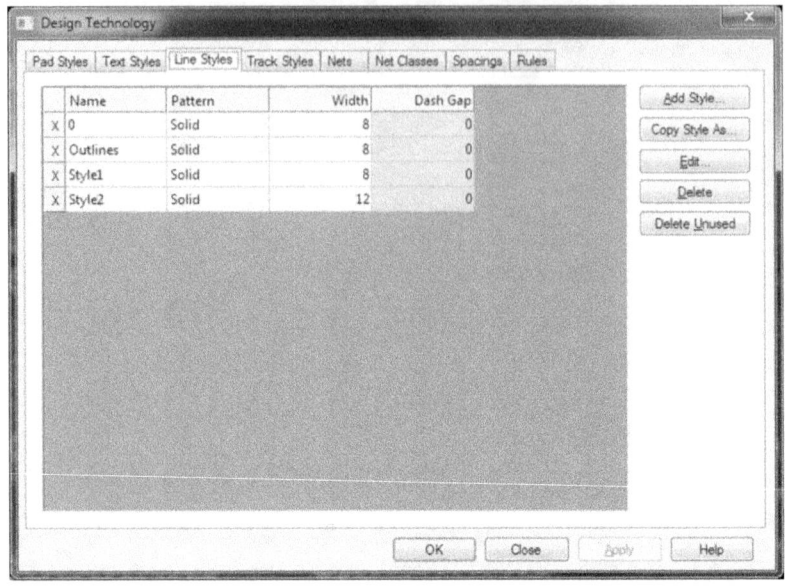

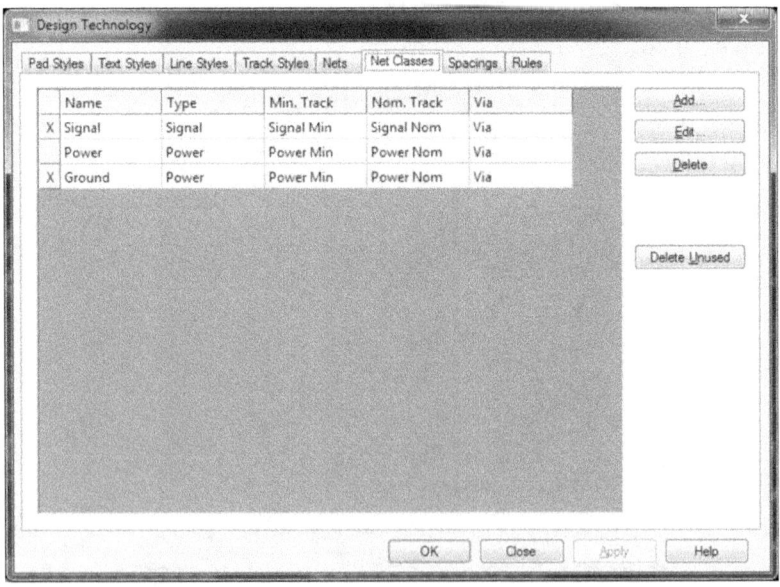

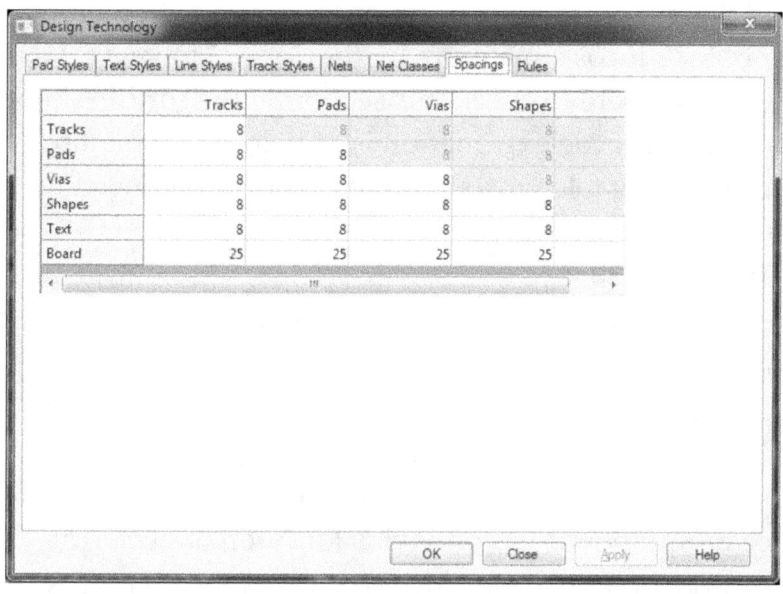

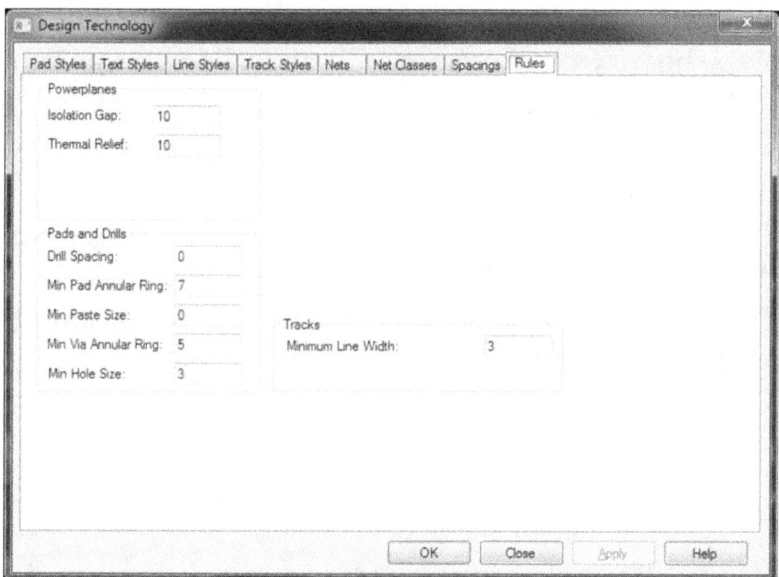

And, in answer to your question, no I don't get paid by the page. In fact, it costs me more by the page. However, I expect you'll be back here to check this section in the future. And the purpose of this little explanation is to help you remember that you have it available when you have trouble and need to return to a baseline setup.

Pad Styles

Pad styles are mainly determined at design time of components. There are a few exceptions. One is the via. The purpose of a via is to provide a conductive path between board layers. A thru-hole board has the legs of the components to provide connections through the board. A surface mount board has only vias for interconnectivity.

The problem with vias is they take up a lot of space because of the annular ring. Each via requires a pad that is larger than the hole drilled into the board. For example a standard prototype board from Advanced Circuits allows a minimum 0.006″ line surrounded by a minimum of 0.006″ with no conductor. So three parallel wires would be three conductor widths (6 thou * 3) plus four space widths two between conductors (traces) and two on the outside of each trace for (6 thou * 4) or a total of 0.042″. The via size indicated above is 0.048″ and there would have to be 0.012′ of space around the via for clearance.

To change via size type Alt-S and select Styles. Select the Pad Styles tab. Click on Via and select the Edit button on the right.

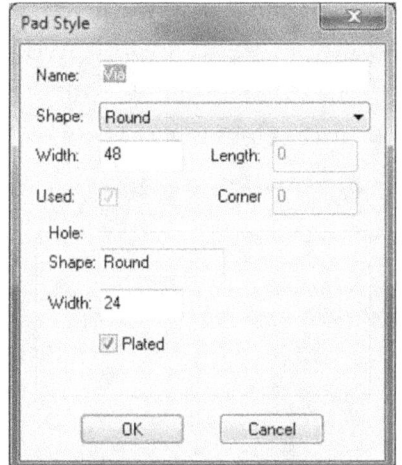 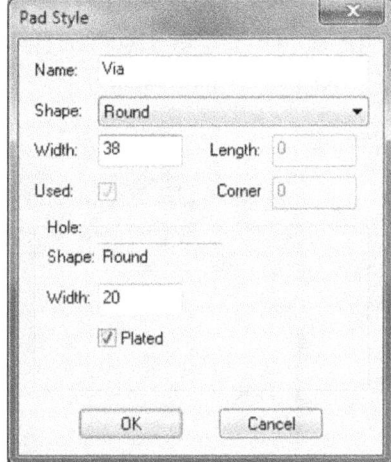

Changing Net Classes

In the schematic editor I changed the names of the two power input nets that are attached to pins PL1 pin 1 and PL1 pin 4 as PL1-1 and PL1-4 respectively. Type Alt-S and then N. Select Delete Unused. The Net GND is already denoted as ground. PL1-4 is also carries ground current and should ave a class of Ground. Select PL1-4 and select Edit. Select Ground from the Net Class: pull-down menu and select OK.

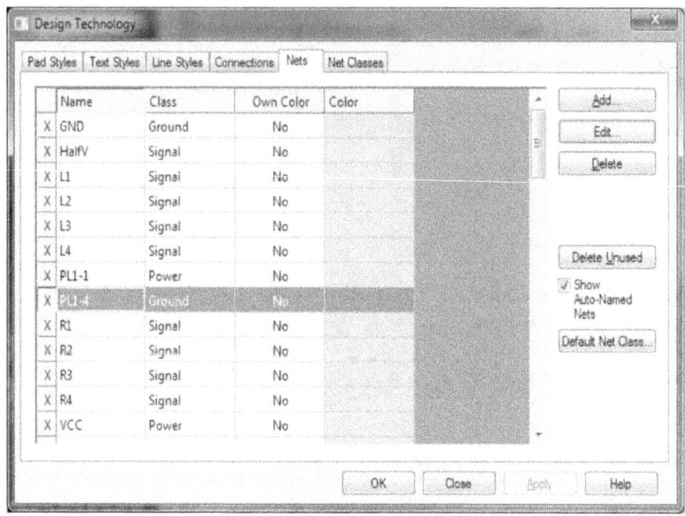

Set Vcc, GND, PL1-1 and PL1-4 according to the following screen.

Updating PCB from Schematic

We made a change in the Schematic and we need that change to be reflected in the PCB artwork. Press Cnt-S to save the design. Press Alt-T and select select Schematic ↔ PCB to update the PCB with the latest schematic changes.

Returning to the PCB Design: Window we can now manually route power and ground. Press Alt-A then K. and click on PL1-1. Press L and set the New Layer to Bottom Copper with the pull-down menu.

Note: Generally speaking routing a PCB is usually accomplished by routing bottom layer components in one direction, say vertically, and top layer components the other direction, thus horizontally. This reduces the collisions of conductors on a layer. This matrix is loosely followed by PCB auto-routing software and most will attempt to connect components in that manner as the first autoroute pass.

Note: When routing a PCB in uncongested area it is common to wire on the bottom side of the board so connections can be inspected without removing the connector. Connectors are more likely to have force exerted on them and thus are more likely to have open circuit faults as the conducts are broken from the flexing. We do this for the managers and do they appreciate it? No.

I have routed the power supply section. You may notice that even though the pin 1 (lower) ends of R8 and R9 were closer I wired directly to the capacitor. This technique is call a start connection. One should wire from a power source, say the USB connector or a battery, to the on-board storage device which is usually a capacitor(s.) Circuit components should then be wired to the capacitor without having any of the trace common to the input traces. Trust me, this is the correct method to connect external power to on-board power.

And if you don't trust me, a worthwhile tutorial is at:

http://www.learnemc.com/tutorials/Common_Impedance_ Coupling/conducted_coupling.html

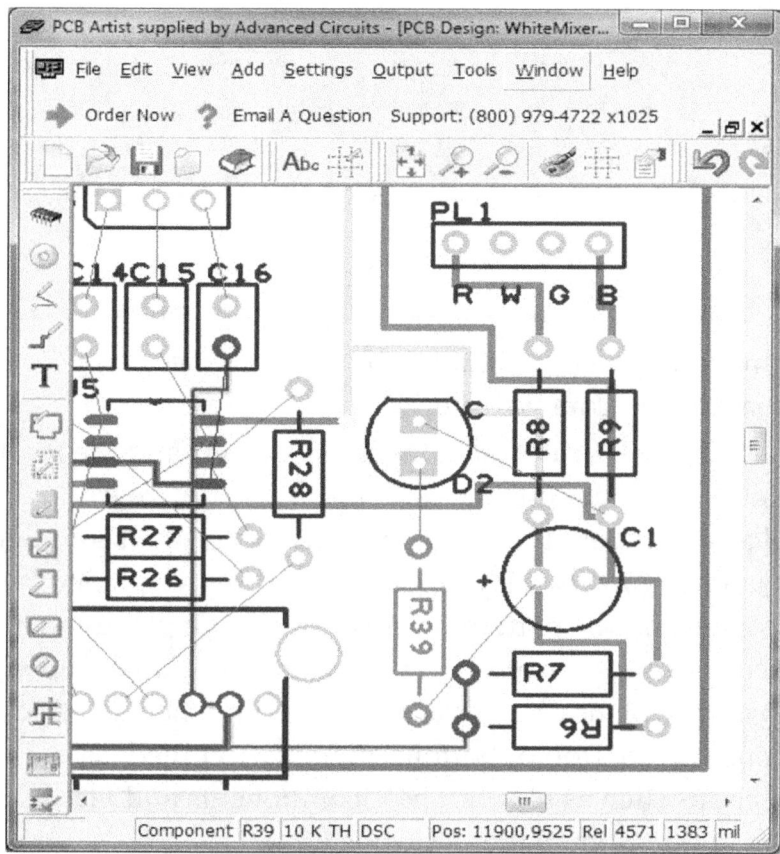

Component R39 10 K TH DSC   Pos: 11900,9525 Rel 4571 1383 mil

Autorouting

An autorouter is a blessing....and a curse.

I've prerouted the power for the PC board because it is an analog circuit. I also made some other preroutes that would be difficult for the router to resolve.

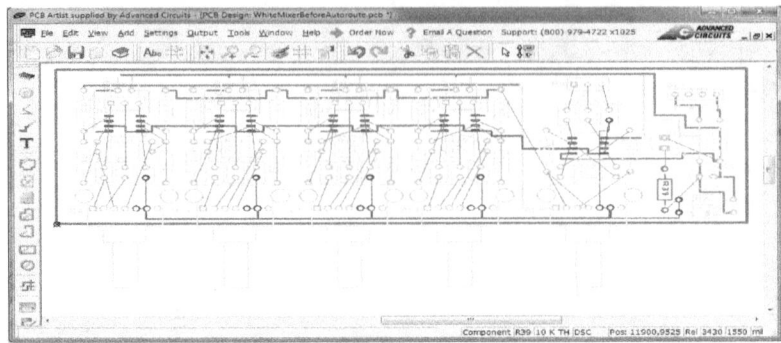

Remember you can get the pcb files from the AbrahamAaron.com website.

Select Tools → Autoroute - Nets → All Nets and the Route All Nets screen will pop up. The routing software treats the board as a number of planes subdivided into small subplanes. For example, for a two layer board there are two planes the top layer plane and the bottom layer plane. Those planes are then divided into a matrix of x and y coordinates. Connections occupy these planes for surface mount components and routes only one surface (top or bottom) plane is occupied. For connections such as through hole components and vias both the top and bottom planes are affected.

In the process of attempting to route the board the autorouter takes note of all the pins connected to a node. The locations of those pins are noted for each active plane of the board. These locations may no longer be utilized.

After all the components are placed the routing software attempts to traverse the unused areas to find a continuous route between pins.

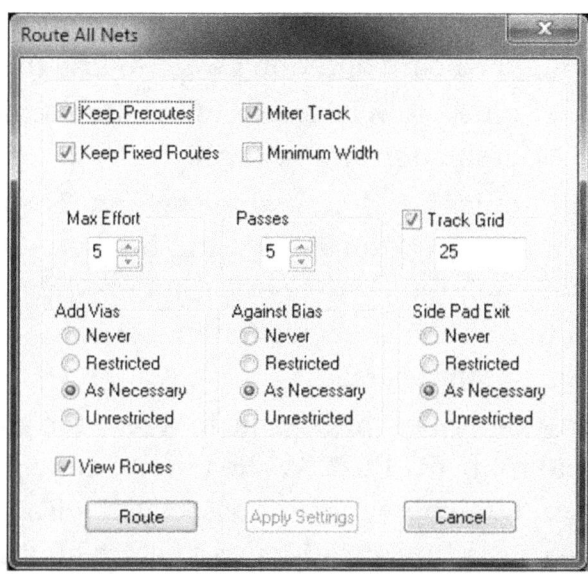

Track Grid sets the width of those routes. As the width increases the number of available areas for routing decreases but routing speed increases.

I have found in practice that having a large Track Grid for initial routing attempts results in boards that have more possibilities for either manual or automatic re-route.

The signal trace width for this board was initially set at 0.008" and with a track grid of 50 thou that is six track widths. Another special thing about a 0.050" Track Grid is that many through hole components such as our capacitors and connectors are on .1" centers and 50 thou allows a line between two component pins. If the board doesn't route we can decrease this value and gain more routes or more easily work with the result should we decide to manually route.

The pots are spaced more closely but they are placed at an edge of the board so that all the pins can come off one side of the component.

Keep Preroutes - is checked because we have spent some time prerouting the board and don't want to lose that.

Keep Fixed Routes - can be essential for situations such as RF circuits, EMI reduction and so forth.

Keep Fixed Routes is a double sided sword. The autorouter can do some unexpected things with this box set. After a bit of experience you can predict some of a routers output and take advantage of this option to coerce the router into accommodating your desires..

Max Effort – describes the lengths to which the autorouter will attempt to route the PCB. While I don't know what the spinner selects, I do know that it is associated with the number of alternative routes that are attempted before moving on to the next net.

The board can get pretty wild and a bit more wild with each iteration. No offense guys.

Associated with the Max Effort selector is the Add Vias selection radio buttons. Adding a via can allow you to change sides of the board for a route and then return to the original side for the purpose of avoiding obstacles. However, a via is much wider than a track and will clog up potential routing areas. In tight boards "restricted" and "never" are good choices for repeated routing.

Passes - determines the number of times the router saves its output and begins the process again. This save and restart removes all the accumulated routing attempt information, clears the matrices and attempts to apply the net to the board.

Associated with the Passes spinner is the Against Bias radio buttons. In the not-so-good-old-days when PC boards were laid out pre-photographically with tools such as lined Mylar, pre-printed models and opaque tape, it was standard practice to select a direction of travel for each side of a board. This allowed relatively unrestricted access for long routes without the over and under of spaghetti routing.

In this board the bias is green for the horizontal lines and red for the vertical lines. The longer a layout technician could maintain this vertical and horizontal bias to the layout the more easily the work would be. As you can see here, this is all to be taken with a grain of salt. The Against Bias radio buttons determine how tightly the bias rules are enforced. Or the number of grains of salt to be applied if we follow the metaphor.

Side Pad Exit – is not really associated with any of the other items in the autorouter, but it does affect the other items. Surface mount ICs cover the PC board and are small. If traces come off the IC willy nilly then there is increased chances of solder shorts and board real estate lost due to lost route area close to the IC. This is exacerbated by having defects hidden by the IC.

We want to avoid traces under any IC so manufacturing defects can be found, but this was the exception. A single line straight across with no vias was a small price to pay for loosing two channels (because of the vias) on each side of the board.

The wire exiting the top from pin four is ground for the IC and must be routed to all the ICs.

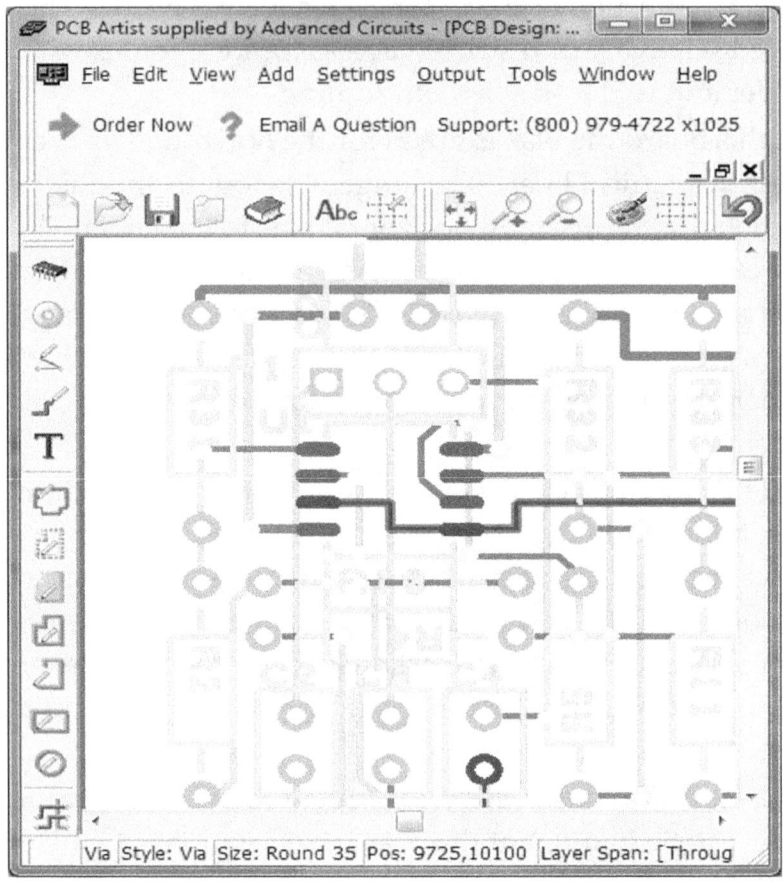

Now that I have a hot-rod computer I always view routes. It makes me feel powerful and omnipotent until it finishes and then I just have to go back to work. Besides, someone might walk by and be impressed.

I had some wonderful screens and repair tips written for version 1.5. But when I installed Version 2.0 the circuit just routed. And I must tell you the route came out quite nice.

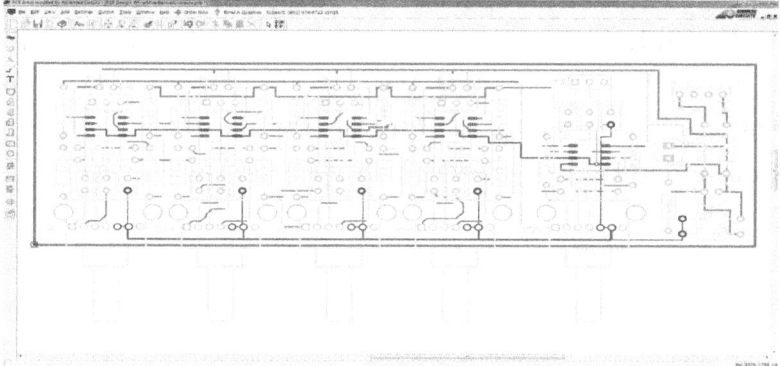

I've hidden the silkscreen. So that I may have an unfettered view of the solder layers. To do this you type Alt-V then C.

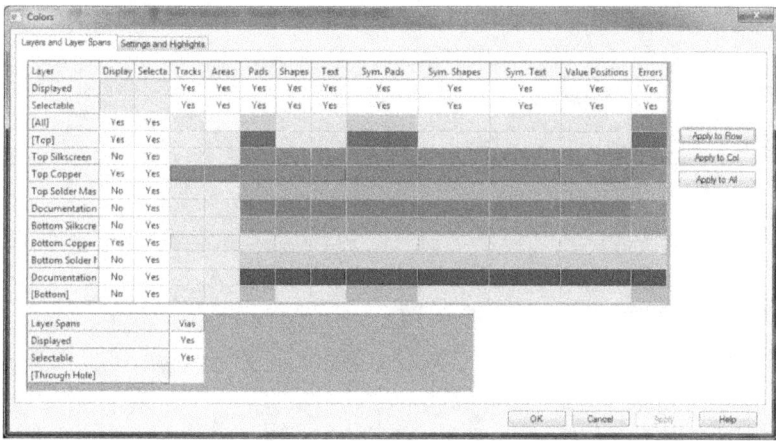

You change what is displayed in the second column from the left of the large matrix. The settings here will determine how components and tracks are displayed.

Post Autoroute Cleanup

First of all, let me say that the online tutorials are excellent for these operations. In fact, they are the best in the business. I suggest that you watch them as I do not have the skill to describe mouse use during rework.

I do suggest that you perform a design rule check quite often. The light colors I use here are not optimum for PCB work, but even with the dark background it is easy to overlay two routes. That isn't too bad unless you base subsequent work on an earlier mistake.

But Wait!

One telling characteristic of world class software is the ability to make changes to the printed circuit board and have those changes be propagated back to the schematic.

In this case we are making this set of changes for demonstration purposes. I would make the change forward (from schematic to PCB) with a simple board that fully autoroutes on the first pass.

This would be the method of choice for a very complicated PC board that required a lot of hand routing. Other reasons for this type change might be to swap pins on a microprocessor design in software instead of hardware due to a difficult routing situation or to make a quick change in something like a connector without having to fully respin the board.

You are going to think that the situation is made up because there is plenty of room in just the right places. You're right!

Your mission Mister Phelps, should you decide to take it, is to place four bypass capacitors on the printed circuit artwork. Each one will be placed closely to the amplifier for a particular channel.

The first step is to place the four capacitors near the board. Type F8 to open the component dialogue box and select the 10uF 6.3 Cer capacitor from the Mixer library. Select the Add button and the component is 'stuck' to your mouse. Place the component by positioning it and clicking.

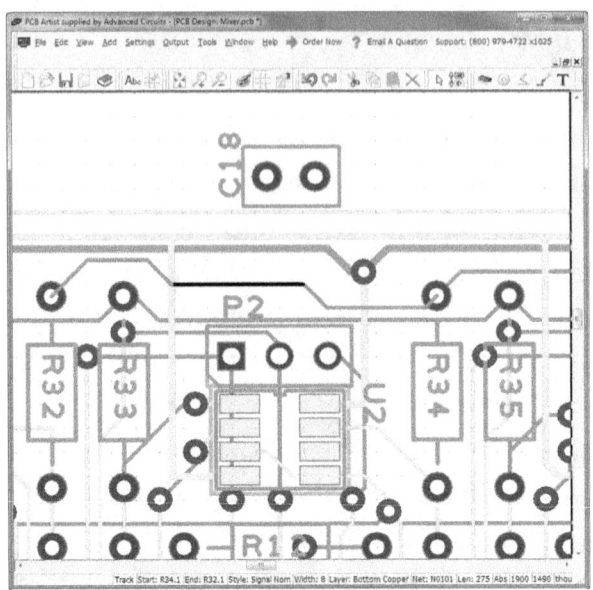

Then use your mouse to zoom into the area where you need to add the capacitor. I've selected the second channel for demo purposes.

I've highlighted the trace I want to move to make room for the capacitor C18 which is now outside the board outline.

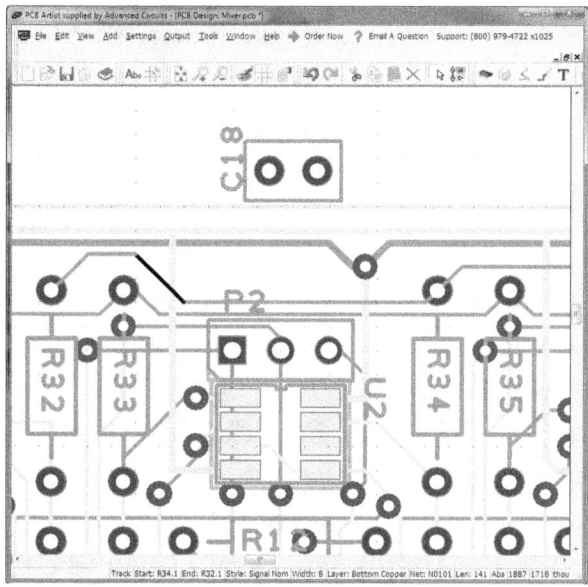

Move the capacitor into the desired space and place it (left click).

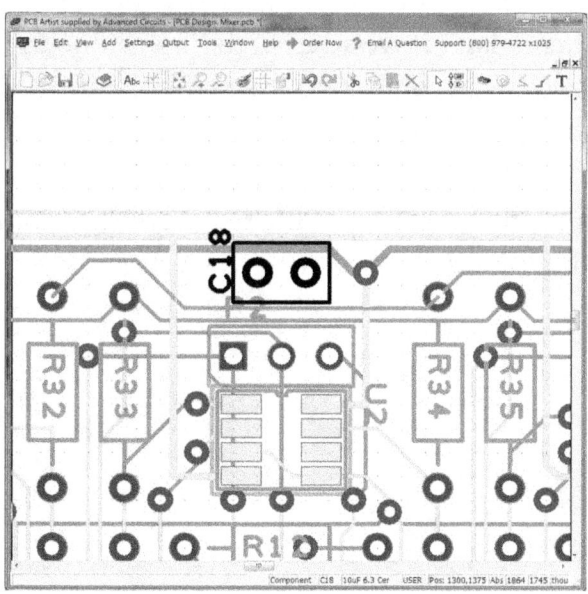

Type Alt-A then k to add a track. Notice that I've added the track directly from the power line to the capacitor. The track style is set by the first connection made and this is a power track so it is important that you click power first.

You should also notice that I have routed the power from the IC's power pin directly to the capacitor for both power lines. This is another example of a 'star' connection in which all connections are directly to the power source. In this case C18 is the local power source.

Now do an Alt -> T then R for a design rule check. Fix 'em if you've gott'em.

Check the AbrahamAaron.com web site for a book regarding routing EMI, RFI, Power and Ground circuits. (Yes that was a shameless plug, but the books are cheap – and inexpensive.)

All the other capacitors are inserted in a similar manner.

The next step is to add the components back to the schematic sheet. Isn't it wonderful to have a one to one correspondence between schematic symbol and PCB symbol? There is never any question as to which schematic or PCB symbol you're editing.

So return to the schematic editor and add the components as described before.

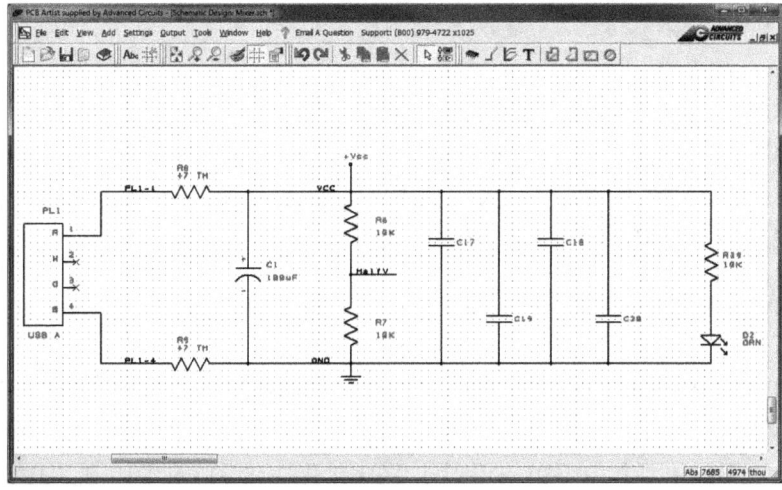

Type an Alt → T and select forward design changes.

I am first and foremost a circuit designer. A schematic capture and printed circuit board program is just another tool in the arsenal. There are other packages that give the capability to create SPICE schematics and have built-in simulators for board characteristics and 3-D drawing outputs for the CAD system. If you need those things then you should examine other packages. I suggest that you consider Pulsonix which is handled by Advanced Circuits and can be found at

http://www.4pcb.com/pulsonix/

Please give any feedback you feel is appropriate to the publisher's website http://www.AbrahamAaron.com .

If you find an error or have a suggestion that is implemented into future editions of this book you will get a copy of that future edition on a first come first serve basis.

If you desire evaluations of circuit design or printed circuit design you can contact the author at www.Aanoit.com (pronounced an-wah which is derived from a Finnish word for provider.)

Let's Spice It UP!

I know it's Korney. I am from the land of Hee Haw.

Spice simulation is one the most significant cost saving innovations in Electronics. It also allows many new ideas to be developed with reduced effort and cost.

SPICE History

Simulators are used in many applications. Possibly the best know simulators are the commercial aircraft simulators that are used when training pilots, making our lives much, much safer in the air. These simulators allow pilots to practice the routines of normal flying without risk to themselves, the passengers or financial loss to the airline. The latter being the motivation for having so many high quality simulators.

These simulators also allow pilots to experience unusual circumstances without risk. One example is if a rudder gets stuck in catastrophic position. The simulators provided the information to point to the problem and effect a design change. Aircraft simulators now routinely simulate this and many other malfunctions to acclimate pilots in dealing with this and similar situations..

SPICE is an electronics simulator. In 2011 SPICE was hailed as an Institute of Electrical and Electronics Engineers (IEEE) milestone along with the discovery of batteries, Alternating Current, Radiotelephone, the transistor and so forth.

When presented with a properly formatted netlist, Spice will very accurately simulate a circuit in minutes instead of the weeks that would be required if accomplished by physical circuit creation of creating, populating and troubleshooting a traditional printed circuit board. Changes and the testing of those changes can then be accomplished in minutes instead of days or weeks.

SPICE was created by Laurence Nagel under the tutelage of Professor Donald Peterson and based on the work of Ronald Roher, all of the University of California, Berkeley. Originally written in FORTRAN, the engineering language of the time, it was converted to C and placed in the public domain.

There are a number of Spice variants with the primary distinctive factor being the user interface. I will describe the use of PCB Artist and LTSpice, a very excellent variant, provided free of charge from Linear Technology.

Linear Technology

Linear Technology, the company, was founded in 1981, and became a force in the development of linear integrated circuits. Their innovative approaches provided products of superior technology and reasonable prices.

LTSpice IV

LTSpice is complete in that it has its own schematic capture software. However, it does not have printed circuit generation software. It would be very inefficient to enter a circuit in LTSpice for simulation and then enter it again in PCB Artist so the interface provided by PCBArtist in version 2 is highly appreciated.

LTSpice IV can be downloaded at:

http://www.linear.com/designtools/software/

How It Works

Generally SPICE reads in a network, makes a few starting assumptions and begins computing the state (voltages, currents and power stored) of a circuit. Assuming the program can reach a steady state at time 0 solution, the program then increments its internal clock (set [controlled] by you); takes note of new conditions (such as a voltage source increasing in magnitude) and computes the circuit's state for the new circumstances. This process is iterated until either a particular state is reached or the timer counter (again set (controlled) by you) reaches terminus[6]. The process is quite simple in concept, but difficult in execution.

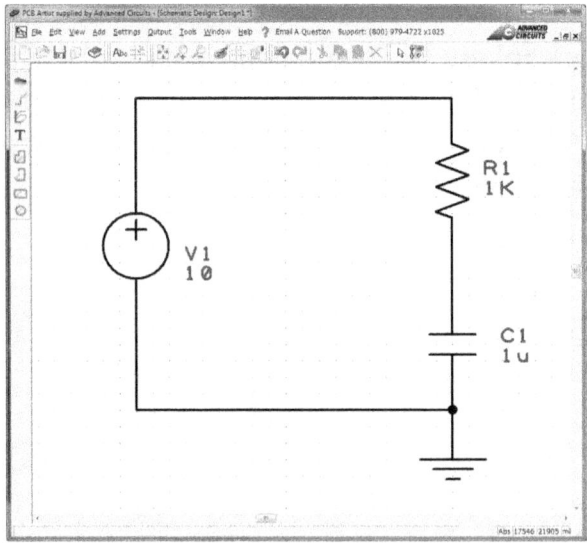

---

[6] I have waited for years to use this word in a sentence – I could have said endpoint, but I wanted to use terrminus. Now, I've used it twice. Another word of the day is implemented.

Limitations of SPICE

The root problem in using SPICE is "How close is close enough?" The simple, on the face of it, circuit below[7] would seem to be easily simulated.

In fact, a capacitor is one of the more difficult items to simulate. The current through the capacitor depends on the applied voltage less the voltage drop across the resistor. Thus the resistor's current and the capacitor's current result in a 2nd order equation. The combination of a resistor and capacitor results in a equation that involves a natural logarithm to solve.

As you will see in a later chapter it is difficult to express the outputs in graphic form as accurately as one might think due to the difficulties of plotting a curve on a digital device.

This is not a knock on LTSpice or PCBArtist it is simply the nature of SPICE itself.

---

[7] I'll demonstrate how to input that circuit later in this book.

Your First Simulation

Install LTSpice.

It can always be found at www.linear.com but for now it resides at:

http://www.linear.com/designtools/software/#LTSpice

You can either download the software to a location of your choosing or install directly. I prefer to get a copy and install from that local copy. Sometimes companies decide to charge for their software and sometimes they change the software with the second copy not working as well as the first. I like being able to re-install the older copy.

Installation is straight-forward on my 64bit Windows 7[8] machine.

Creating a schematic in LTSpice

Why the dickens would you want to create a schematic in LTSpice when we are wanting to work directly from PCB Artist? Because it is instructional. We will compare the models from PCB Artist and LTSpice.

Start LTSpice.

Type Alt → F then N. You should be editing a schematic probably called Draft1.asc.

Press F2. You should get a screen similar to the following screen. Using the horizontal location bar or the right keyboard arrow key move to the right and select voltage. Select OK and you should get a copy of the voltage source schematic model in the schematic.

---

[8]And I don't expect to upgrade in the near future. They (you) can troubleshoot Windows 8 without me.

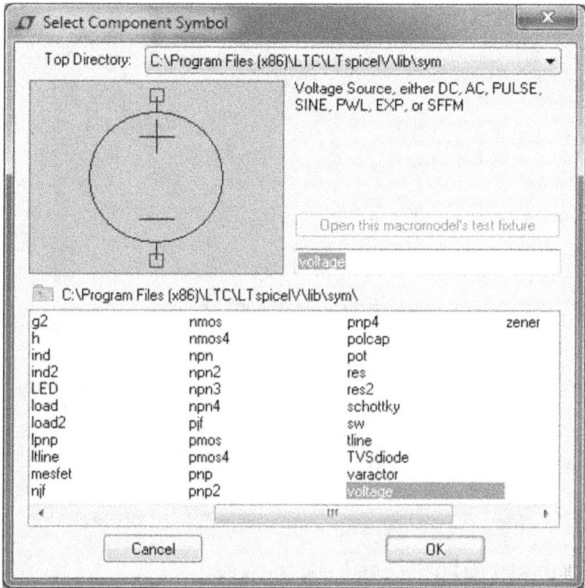

You must then set the Voltage and Series Resistance to 10 and 12e3 respectively.

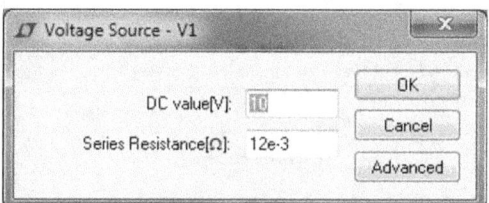

Every real voltage source has a real serial resistance. In electrolytic capacitors and batteries the series resistance is quite significant. Leaving this a blank or making it 0 will give very pretty but overly optimistic results.

Draw the rest of the schematic in a similar manner using cap and res models. Cap should be set as follows:

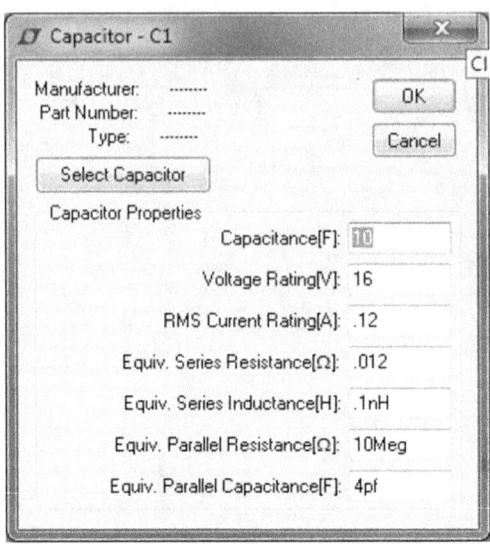

The resistor should be set as follows:

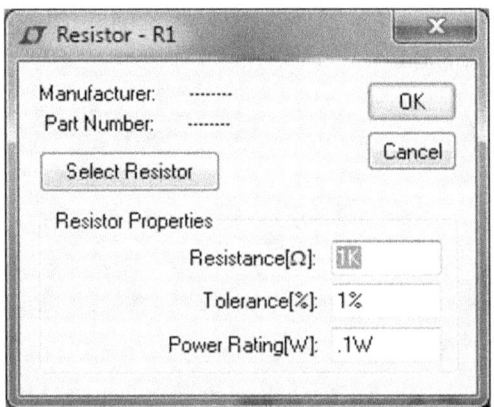

Right click on the vertical line between C1 and R1 and select Label Net. Set the net to Tap. Set the line between R1 and V1 to Vcc by the same method.

You should now have a schematic that looks like:

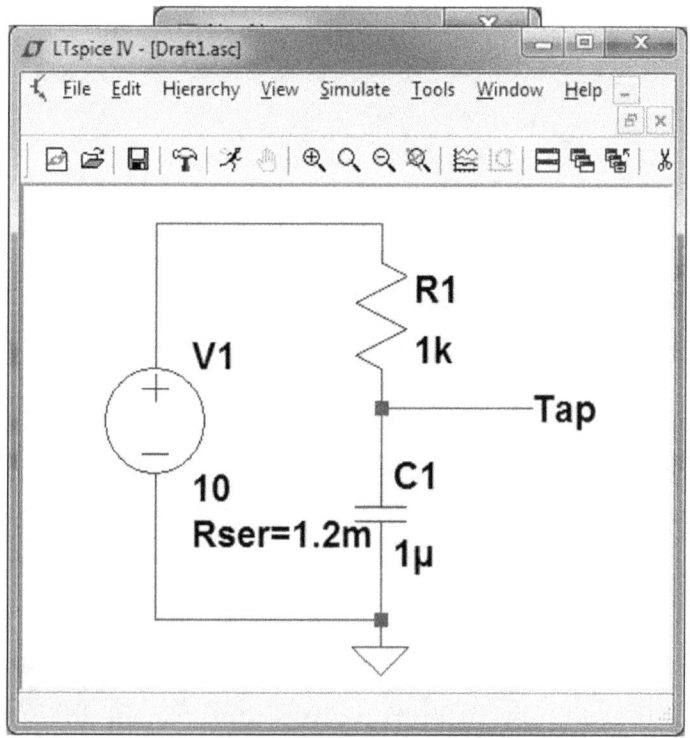

Type Alt-S then select run.

Note that you can always get back to this screen by typing Alt → S then selecting Edit Simulation Command.

When the results first come up you may (probably won't) have a trace on the black screen. The circuit has been analyzed but we haven't provided instructions as to what should be displayed.

Click in the schematic area of the screen (gray background) and type Alt → V then V and select V(n001). This is the node between C1 and R1.

Note: All voltages are assumed to be between the node and ground.

Note: Don't forget to add the ground, I did.

After clicking you should get the following screen:

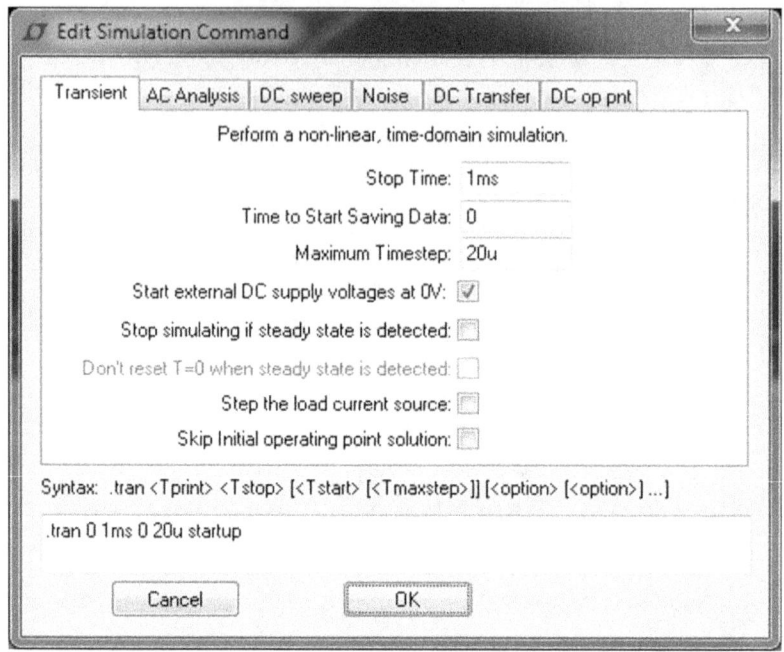

Notice that I have different values than for the PCB Artist circuit in the next section. I do that to show the default presented by PCB Artist and the method to change those defaults.

Also note that I changed the background color of the simulation to white, you will probably get a black background. I'm writing a black and white book.

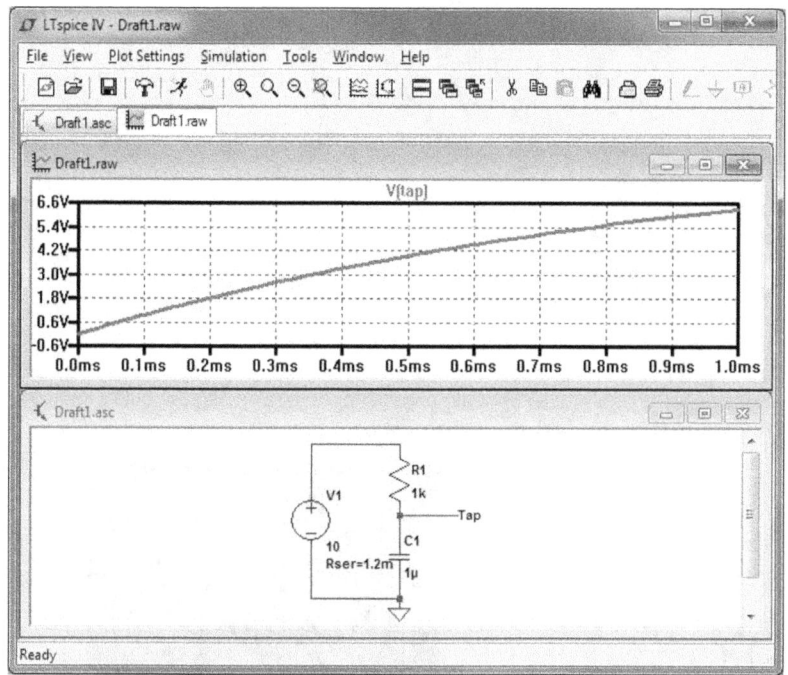

Creating a SPICE schematic in PCB Artist

Open PCB Artist and create a new schematic using Cnt →
F8.

Type F8 to get the Add Component screen.

From the Library: pull-down box select the ltspice.cml
library.

My personal preference is to check the Preview box. On a
slower computer this may significantly increase the display
time.

Using the Component: pull-down box select Power Supply.
Select the Add button and place the Fixed Voltage Source in
the schematic.

In a similar manner place a Resistor (Z shape) and
Capacitor in the circuit. Wire the circuit in series with the
resistor being connected to the + terminal of the voltage source.

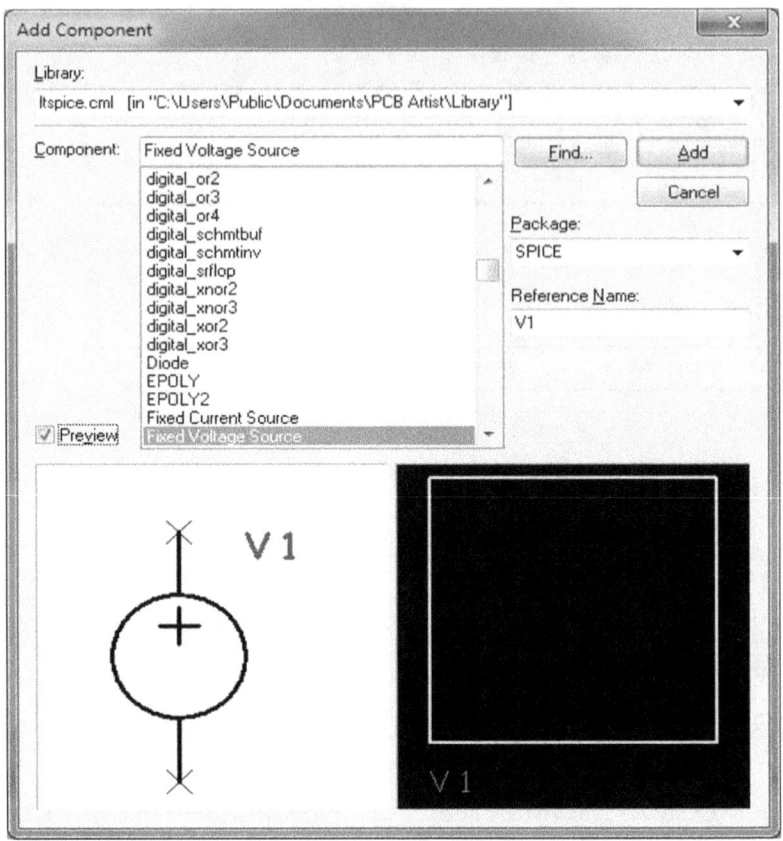

Don't forget to add the ground from the LTSpice component library. All SPICE circuits require a ground somewhere in the circuit. All voltages are measured with respect to this ground.

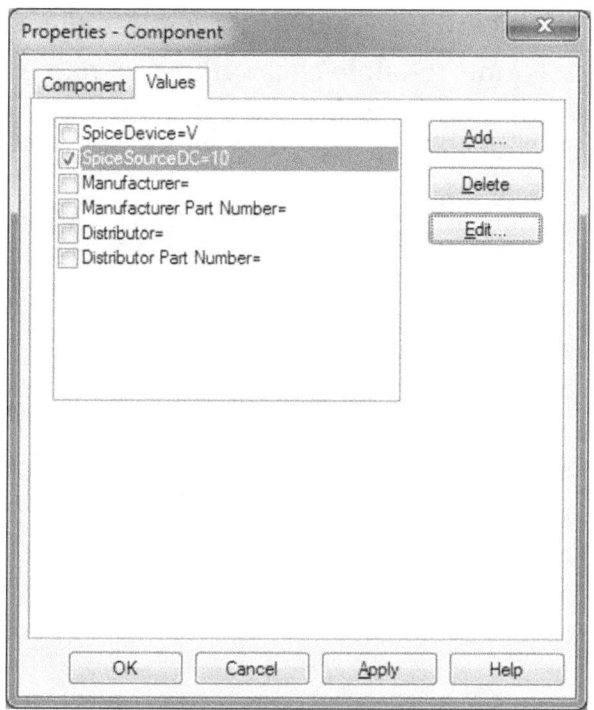

Right click on V1 and select the Values tab.

Double click on SpiceSourceDC and set the value to 10. Spice assumes that the values entered are Volts, Ohms and Farads for voltage sources, resistors and capacitors respectively. Watch out for those Farads! A .001 uF capacitor is denoted by 1n. The n stands for nanoFarad or 1 X 10-9 Farads. See the attachments for common SPICE abbreviations.

Check that the capacitor value is indeed 1n and that the resistor value is 1K.

We can also set up a measurement point by adding a Voltage Probe from the LTSpice library. This is a convenient tool that isn't available from the downloaded LTSpice library out of the box.

Right click on the trace between R1 and C1 and select Display Trace Name. Double click on the net name, probably N0000 and select the Net tab. Type in Junction and select OK. This gives us a point to measure.

Also, never, ever, forget to add a ground to your circuit. It must be the ground from the LTSpice library.

Your schematic should resemble the following:

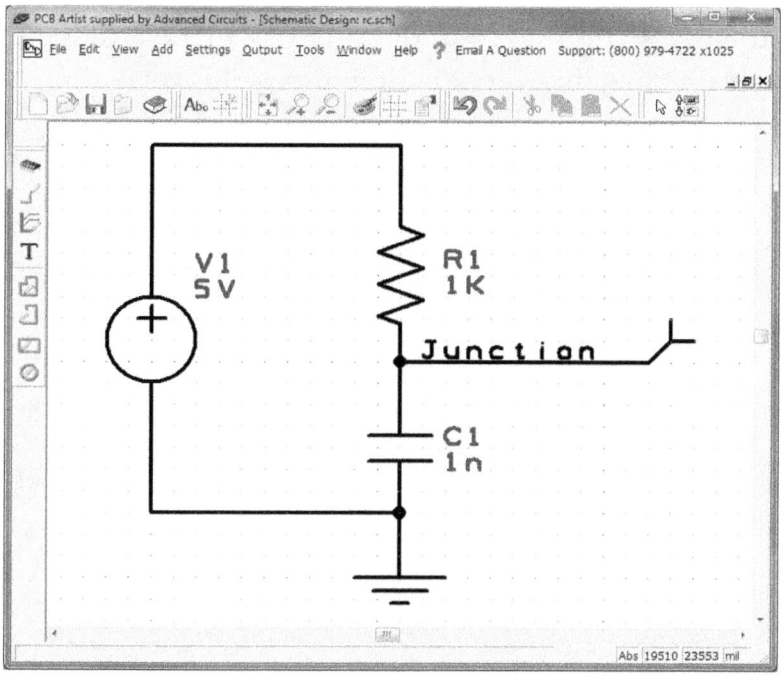

Type Cnt → O and select SpiceSimulationOutput. Select Browse and set the OutputFileName box to a convenient location and name and select OK.

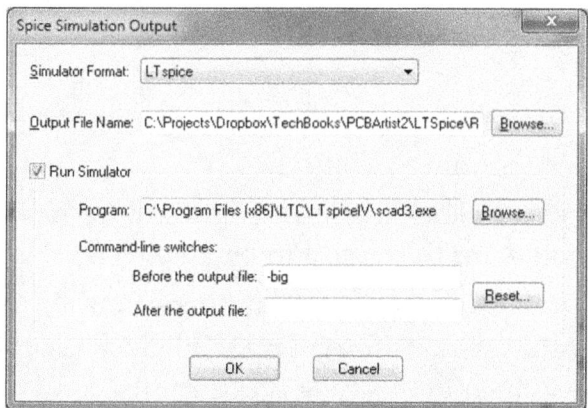

You will also need to set Program. This is the location of the simulator. A few hints are that it will possibly be in the Program Files (x86) or Program Files directory. It will possibly be in an LTC directory under LTSpiceIV and be called scad3.exe.

Select the OK button. This will also save the settings you have made in this screen.

You should get a screen as follows:

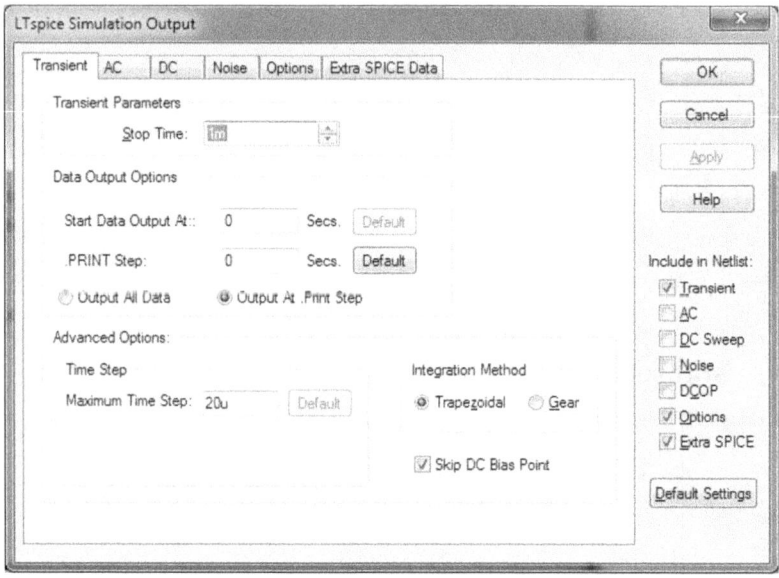

You are now running LTSpice.

This first screen, above, allows you to set the information about how you want the circuit to be analyzed.

You will probably want to be running Transient Analysis. In transient analysis the aforementioned clock is set to time out at Stop Time: which in this case is one millisecond.

Other notable options are Start Data Output in the case you want to see only a window of the total run time and .PRINT Step where you control the output separately from the Time Step, which is discussed below. This will reduce the amount of data storage.

Time Step is merely a suggestion to the simulator. Usually it is better to select the Default button for good performance.

Integration method is almost always best set to Trapezoidal. The exception is when the output will not settle down when you think it should. Think of Gear as having a bit of hysteresis in the calculated values.

Skip DC Operating Point

Ours is one case where this should be checked. As I mentioned earlier the simulator makes a few assumptions before the first calculation. With this box checked the circuit is assumed to have had time to settle into a steady state condition. For example, components like capacitors are assumed to be charged.

However, this is not the case in realistic circuits as the circuit has to reach equilibrium after power has been cycled from off to on. If we select Skip DC Operating Point the power is assumed to have been off and then turned on.

There is a list of items that are to be included in Netlist: We have selected Transient and Options and Extra SPICE. AC, DC Sweep and DCOP are special types of analysis which we will discuss elsewhere.

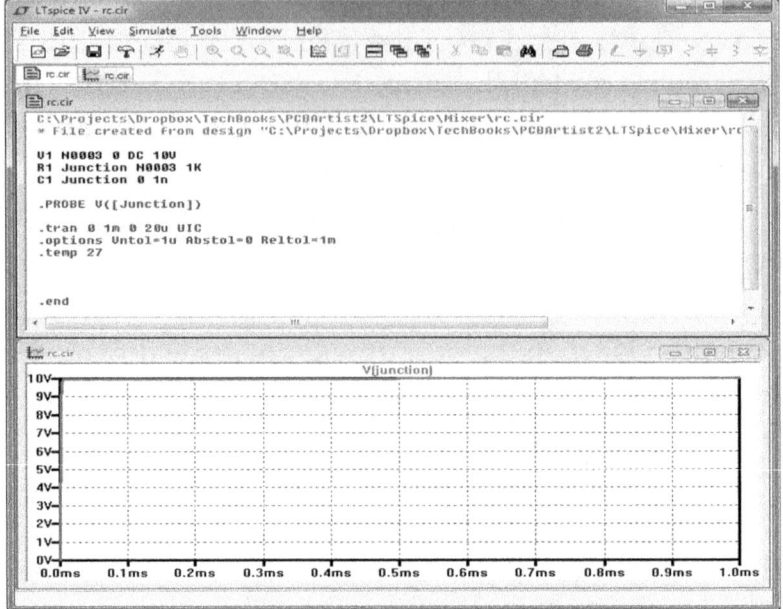

There is not much useable information here.

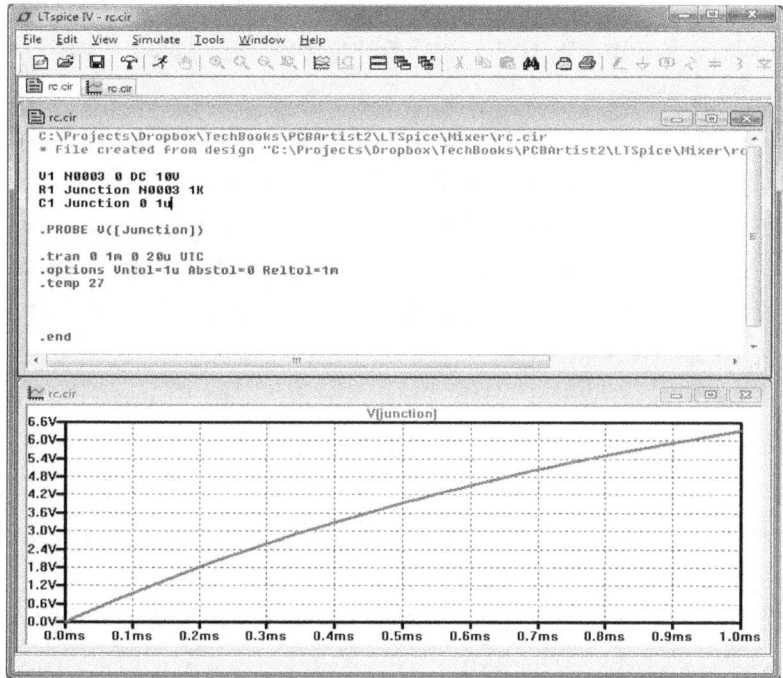

Recalling our electronics a time constant (the time to change 63% of a transient) for a 1K ohm resistor – and one nanoFarad capacitor combination is 1 microsecond.

Let's close out the LTSpice session and return to PCB Artist. Then let's edit C1 to be one microFarad.

Press OK and you should get something like:

That circuit change certainly took less time than finding a new part, warming up the soldering iron, removing the part and retesting didn't it?

The LTSpice Display

LT Spice has presented you with two separate windows. The screen with a black background is the waveform screen and the screen with the white background is the net screen. The waveform screen gives a visual representation of the output from the most recent SPICE run. The net screen lists the net being used with the lines beginning V1, C1 and R1. You read it as:

The component V1 that is connected between node N0001 and node 0 (Ground) of the type DC with a value of 10 and the component C1 that is connected between node Junction and 0 (Gnd) with a value of 1u (1 microFarad) and the component R1 that is connected between Junction and Net N0001 and has a value of 1K (1,000 ohms.)

The lines beginning with .tran, .options and .temp are the simulator settings. You read it as:

Perform a transient analysis with an output step at a SPICE determined interval and from 0 to 1 millisecond(s) with steps between 0 and 20 microseconds beginning with Initial Condition (no quiescent circuit analysis before beginning.)

Simulate with a Vntol of 1 microVolt, an Abstol of 1 picoVolt and a RelTol of 0.1%.

Simulate at a temperature of 27 degrees C.

RelTol is relative to the signal levels at that point in time. , Vntol and Abstol set the minimum voltages in an absolute manner. In short, these are pretty good values, but may be changed to achieve a quicker resolution and slight increases in accuracy.

.end indicates the end of the Spice instructions section of the code.

I implied earlier that it simply wasn't that simple. For example, what if we were wanting to determine how long it took for the capacitor to charge to within 1% of its maximum value?

In our extremely simple example we can tell that the capacitor will charge to very nearly 10 Volts, but circuit theory tells us that it will never, ever get there. If we run the simulation twice we can get the answer. We can run it first with the SkipDC Bias Point box checked which will give us the final charge voltage. We can then deselect the Skip DC Bias Point unchecked and run it again to find the point on the curve where the capacitor has sufficiently charged as measured at Junction.

SPICE Measurements

The first run tells that the capacitor will charge and that is 10Volts. We then multiply 10 Volts by .99 to get the 9.9 Volts as the final voltage of interest.

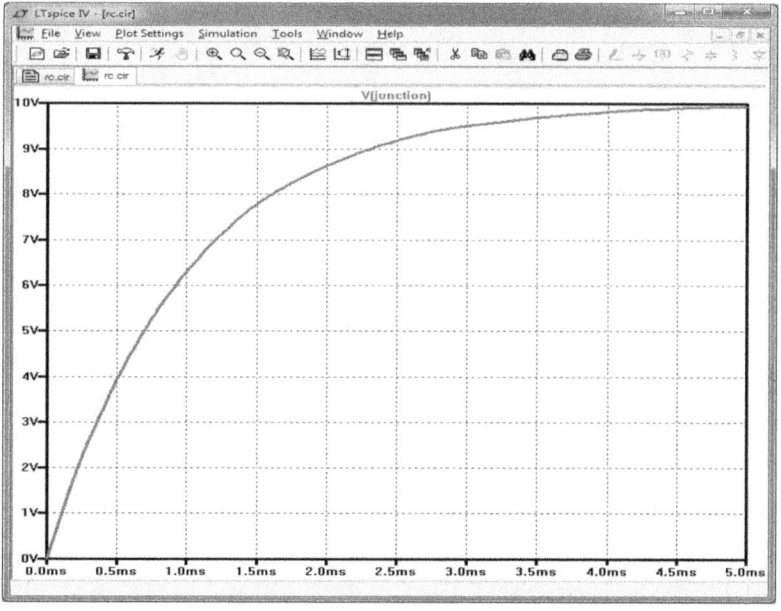

So just where is 99%?

I've adjusted the total run time to 5 milliseconds. In the LTSpice netlist window (the lower window in the display above, I changed the SPICE transient directive from .tran 0 5m 0 20u UIC to .tran 0 5m 0 20u UIC and selected run.

As you can see it is difficult to determine at what time the voltages across the capacitor becomes 9.9 Volts.

The time is calculated as -ln(0.01)^9 or 4.605 milliseconds. But it would be difficult to determine from this plot. To help with the analysis you might double click on the y (vertical) axis labels and set the range to be 9.8 Volts to 10 Volts.

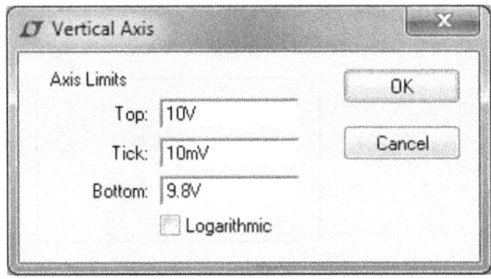

Then double click on the x (horizontal) axis labels and set the range from 4.4ms to 4.7ms.

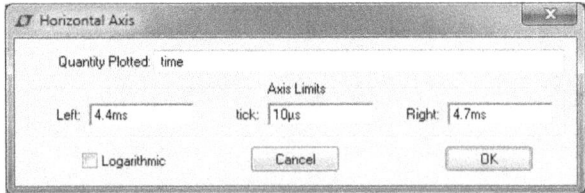

Note that the two displays are equivalent. They made the vertical axis dialogue box vertical and the horizontal axis dialogue box horizontal.... too cool!

This gives us the following display.

---

[9] This is the result of Ln((Input Voltage – Sample Voltage)/Input Voltage)

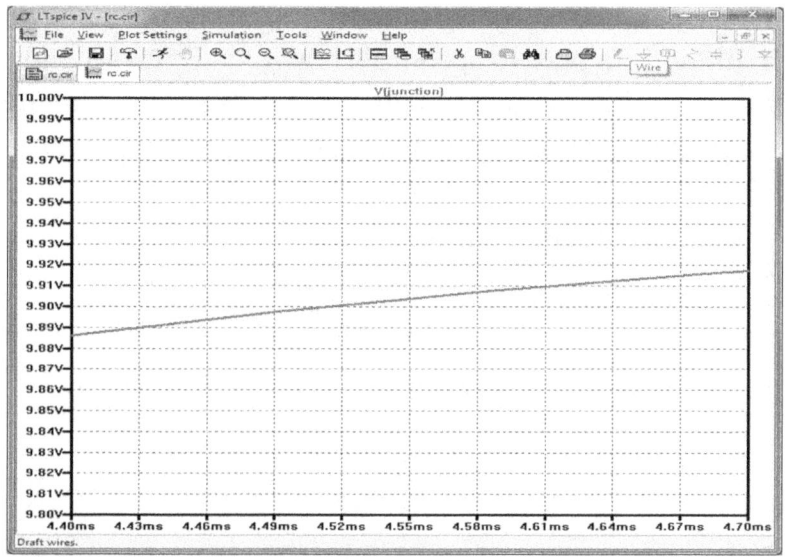

As I mentioned before we know that 99% of a 1 millisecond time constant circuit is 4.605 milliseconds. The value given here is 4.51. There can be a number of reasons for this discrepancy. For the curious type Alt → tools and select the Compression tab and then the Drafting Options tab.

The conversions are made and then transferred to the plotting software and rounding errors can occur. This measurement is on a very flat (the changes in y are small for changes of x) portion of the curve and the changes are nearly the size of pixels.

The time error is on the order of 2% and the voltage error is on the order of 0.08%. So, be careful with this tool and always check the results. It is possible to access the raw data files but that is out of the scope of this book.

Of course I cheated. I had a point to make. Through experience I knew one of the worst cases for any SPICE or electronic simulation program is the exponential charge. My primary impetus was to demonstrate methods for changing the circuit and display to accomplish a goal. But a secondary reason was to show that users often get lulled into believing Spice is perfect. It is really, really, really good; but it's not perfect.

The actual calculations are awesome, but, unless you go directly to the data you are viewing an approximation.

I suggest keeping a scientific calculator at hand.

Updating Version 1.5 components for Spice

We've established that Spice depends on models. Some models such as resistors and capacitors are 'built in' and the models are available by simply typing the predefined identifier.

When you are modifying your existing components it is easiest to use the library manager and examine how information is passed from PCB Artist to LTSpice. SPICE models are simplified by utilizing SPICE Devices. A SPICE device is a generic model of a generic component.

Updating Standard Devices

The two most ubiquitous models are the resistor and capacitor since they seem to include themselves in any electronic design. They are also defined in the base package, meaning you don't have to load a library or subroutine to utilize their models.

This is true of a number of components. One must examine the literature of each implementation of SPICE to determine the models supported for that particular variety.

The Resistor

Let's say we want to make one of our existing resistors LTSpice compatible. Type Cnt → L and select Library ltspice. Then select Resistor(Z shape) and select Edit. Then press Alt → Enter and select the Values tab.

Note that the first two lines are new. Spice Device shows that a standard device, R, is being used. There is another value R that is set to 1K. With respect to creating the SPICE netlist

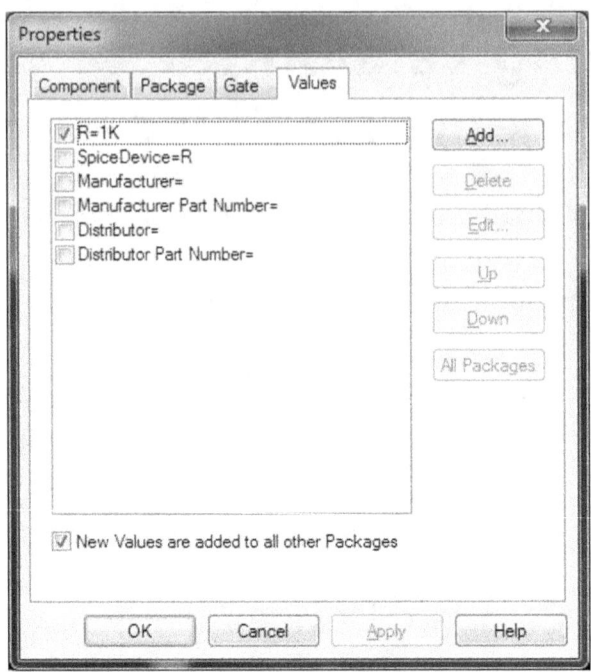

no other value for resistance is significant. No matter what value is otherwise available with this particular device the SPICE resistance is 1K.

But we want more!

A SPICE model may be easy or complicated depending on the desired goal. For most simulations, just the resistance value is quite enough. However, if performance over temperature is important you may change the temperature of components and determine the effects on the circuit. If we look at the default LTSpice resistor entry screen for a 10K resistor similar to the one we are using in the mixer circuit, we find the following:

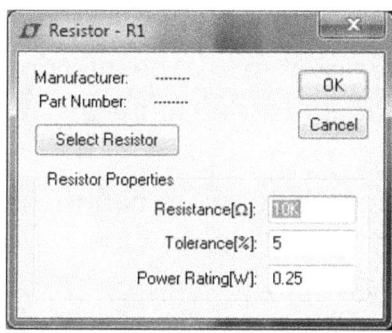

Even though this is the default entry more information may be added, depending on the SPICE interpreter.

To see how this is passed on to LTSpice's interpreter we look at the circuit file with the .asc suffix. And the information is passed in the line

```
SYMATTR SpiceLine tol=5 pwr=0.25
```

In the real world a change in temperature gives a change in resistance. If there is no temperature coefficient added LTSpice assumes a temperature coefficient of 0ppm/°C.

The temperature coefficient of the 10K resistor we will use is 350 ppm/°C. So, if we have a circuit in which the resistor heats up by 20°C due to current heating and is in a 37°C ambient (10 °C more than the specified resistance temperature) the total temperature rise is 30°C. If we multiply the .00035 (350ppm/°C) * 20 we will get a change of 1.05% in temperature. Since this could be either plus or minus 1.05% the total resistance uncertainty is 2.10%. If another resistor in the same circuit is being used as a voltage divider and its change is positive while this resistor's change is negative the total change could be as much as 4.2%.

So if we want to include the temperature coefficient in our model the R value would be

```
10K TC1=350u
```

I like using u because it forces the interpreter to express the values in a more readable form. To me 350u is easy to interpret to 350ppm.

We may now combine this information into a single line

`10K TC1 = 350u tol=5 pwr=0.25`

Which we enter as follows.

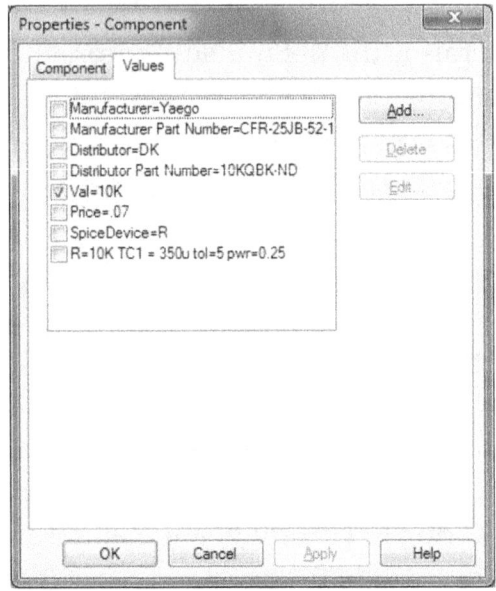

Since I use the total information per part model for my parts library I don't have to be concerned that an 0805 resistor be confused with a through hole resistor or a 1% vs. a 5%. The down side is that I have to add this information to each component in my library(s)...... some day.

The Capacitor

The second most used component is the capacitor. If we look at the generic capacitor model from the LTSpice component library

The two added Values are Spice Device and C. SpiceDevice is the generic model for a capacitor. C is the value that will be passed into the netlist for that C.

Let's look at C1 the first component in the Mixer schematic. Select C1 and right click and select Properties and then select the Values tab.

Remember that this is our highly modified library component. We added a number of fields to create a more robust bill of materials. In addition the capacitor component properties screen above has two extra fields added.

As with the resistors the two items at the end of the component properties are SpiceDevice and C. SpiceDevice tells LTSpice which generic model to use, which, in this case, is C for Capacitor.

The PCB Artist symbol uses an abbreviated model.

Since we are going to plug this into an LTSpice simulator we can peek at their capacitor. Capacitance is denoted as being in Farads. A 100uF capacitor would be `C =100u` .

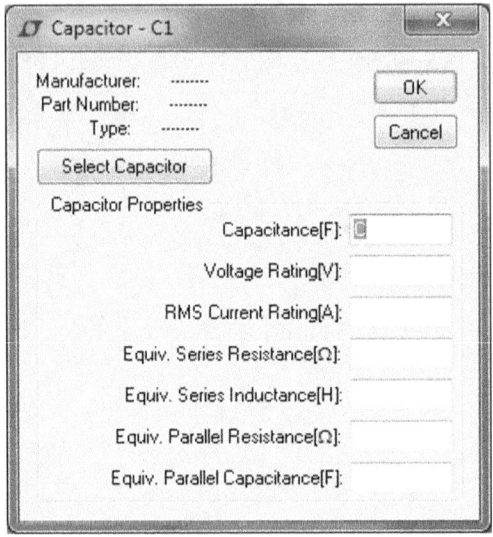

Above is the generic LTSpice capacitor model and its options. It would be a shame to have these extra capabilities go to waste.

We want more!

For the curious, the SPICE expansion of the above is

Property	Unit	Measure	Option Name
Capacitance	F	Farads	default
Max Voltage	V	Volts	V
Max RMS Current	A	Amps	Irms
Equiv. Series Resistance	Ω	Ohms	Rser
Equiv. Series Inductance	H	Henrys	Lser
Equiv. Parallel Resistance	Ω	Ohms	Rpar
Equiv. Parallel Capacitance	F	Farads	Cpar

The LTSpice SPICE output string for our C1 (power supply capacitor) would be:

```
C1 N002 0 100µ V=100 Irms=.2 Rser=2n Lser=4m
Rpar=500meg Cpar=.06n
```

Which reads as "a 100 microfarad, 100Volt, 200milliamp ripple current, with 2×10^{-9} ohms series resistance, 4 milliHenry series Inductance, 500 megOhms parallel (leakage) resistance and 60 pico farads equivalent parallel capacitance."

Oh if only I could purchase this capacitor!

The model screen for that a capacitor from the PCB Artist LTSpice component library would look as follows:

The detail is not there, but for first shot analysis this will be adequate.

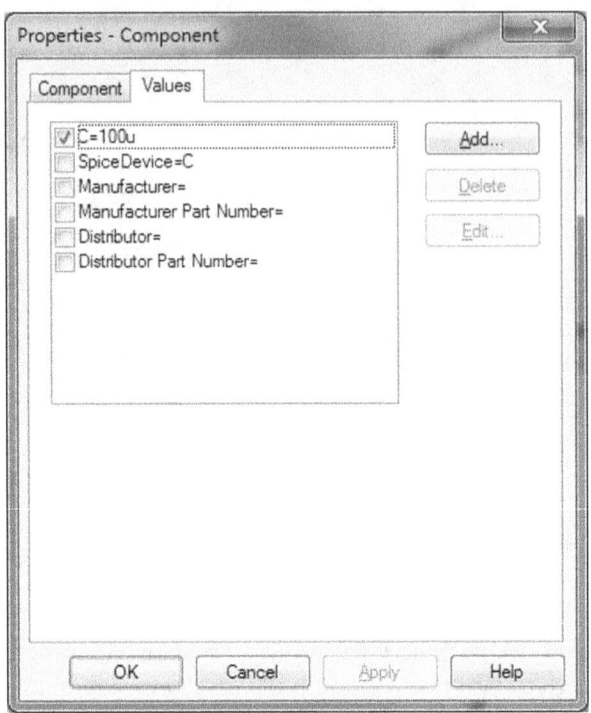

In reality as far as SPICE is concerned the string it receives is:

```
C1 0 VCC 100u
```

This reads as "Component C1 connected between node 0 and node VCC of value 100microFarads."

The point is that there is not a one-to-one correspondence between the models of LTSpice and PCB Artist, but the names have been changed.

With respect to models these are created much like the method we used to add our bill of material information to components. If in doubt about how PCB Artist is handling the interface you can type in Alt → O then S and read the netlist directly. Or you can look at the .cir file that is in your design directory.

The 100uF value is adequate for most applications, however, lets say you wanted to provide a bit more information. Lets say you wanted to model the actual ESH107M016AE3AA capacitor from Kemet. You can download the data sheet at:

http://www.kemet.com/kemet/web/homepage/kechome. nsf/vapubfiles/KEM_A4005_ESH.pdf/ $file/KEM_A4005_ESH.pdf

```
100µ V=16 Irms=.16 Rser=1.8m Lser=4u
Rpar=.27meg Cpar=.06n
```

Rser is a complicated item. Let's assume that this is a 120Hz (the output from a 60Hz bridge rectifier) rectifier circuit (which it is the worst case for most USB sources – a battery operated part would have a lower Rser.)

From the data sheet table the dissipation factor for a 16V 100microFarad ESH series Kemet capacitor is 18.

Capacitor impedance is:
```
1/(Capacitance (.0001) * Frequency (120) * C )
```
or
```
13.2 ohms
```
The Series resistance is therefore:
```
dissipation factor / (6.28 * Capacitor impedance
(13.2))
```
or
```
1.8 milliOhms
```
The parallel resistance is dominated by leakage current. After fully forming the leakage current is .03CV(uA) + 10uA so for our capacitor it is
$$(.03 * .0001 * 16) + 1X10^6$$
or
```
58 microAmps
```
Since this is at 16 Volts the parallel resistance is
```
16/58x10-6
```
or

`270K ohms`

Derivation of the parallel capacitance is left to the reader[10].

Updating an OpAmp

We are going to adapt the Texas Instrument OPA2322 It would certainly be easier to use an LT opamp but that would be contrary to the purpose of the book[11].

Since I have two rather contrary components to develop I have decided to reduce the schematic to its minimum components and still be a working schematic.

---

[10]This is another of those things I always wanted to add to a (semi) serious manuscript. I don't know how to back into this from the data given, but I can tell you that for an electrolytic capacitor this is the non-electrolyte portion of the capacitance between the terminals. You might think of it as the capacitor without the electolyte. The value given is a semi-educated guess (as opposed to a wild-assed guess.)

[11]Which is instruction – not inflating my ego. Well maybe a little bit.

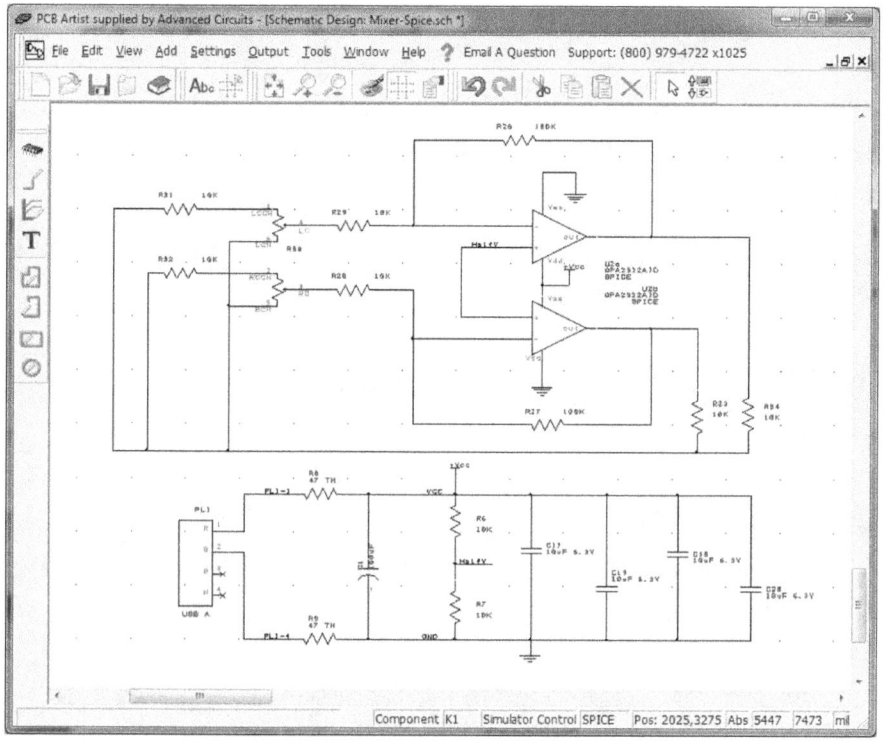

The important things are the circuit performs the original function, an inverting opamp, and I am using both packages of the pot and opamp.

The model for the opamp is a bit easier to obtain than the pot model. Scoff if you will, but it's true.

I am not familiar with SPICE models that can handle two independent circuits in a single symbol but we can use the multiple instantiation method which is to create a single PCB symbol and use two schematic symbols to describe the part.

Creating a Passive SPICE X model

For those items not 'built-in' an external model must be used. These models have an X prefix informing the SPICE interpreter that this component requires special handling.

We will need a new potentiometer symbol. Type Cnt → L to open the library manager

Let's start by creating a new component with the library wizard. Before we can use the Wizard we must create an appropriate schematic symbol.

The original symbol we drew had both pots in a single schematic symbol.

Select Schematic Symbols and then select Edit and then the MixerPot schematic symbol.

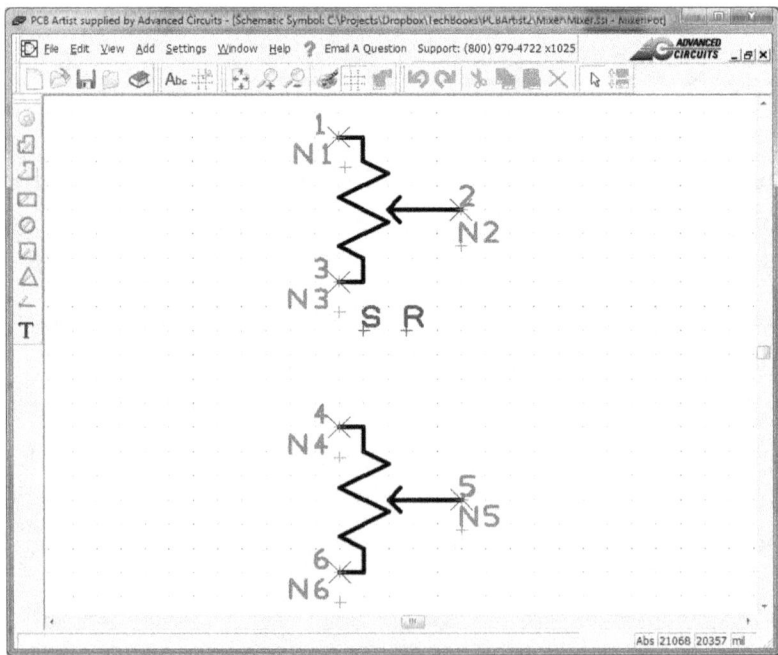

Draw a box around the bottom potentiometer and press delete on the keyboard. Notice that I took this chance to reposition the R (the reference designator location) and the S (the symbol center around which the component will rotate or flip.) I also removed those dreadful ears. They seemed like a good idea at the time.

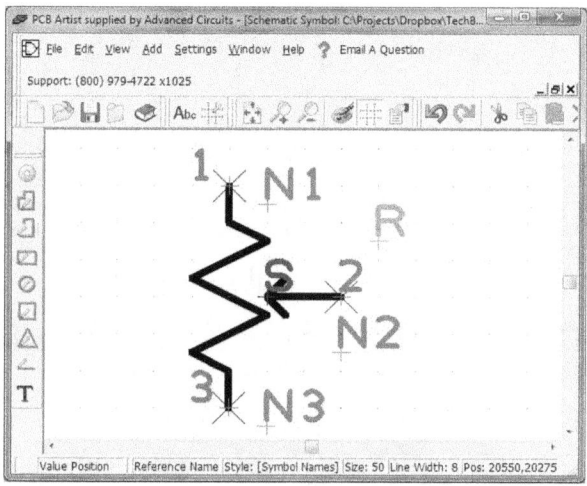

Save the result as NewPot in the mixer library.

Type Cnt → L to return to the library, select the Components tab. Select the Wizard Button.

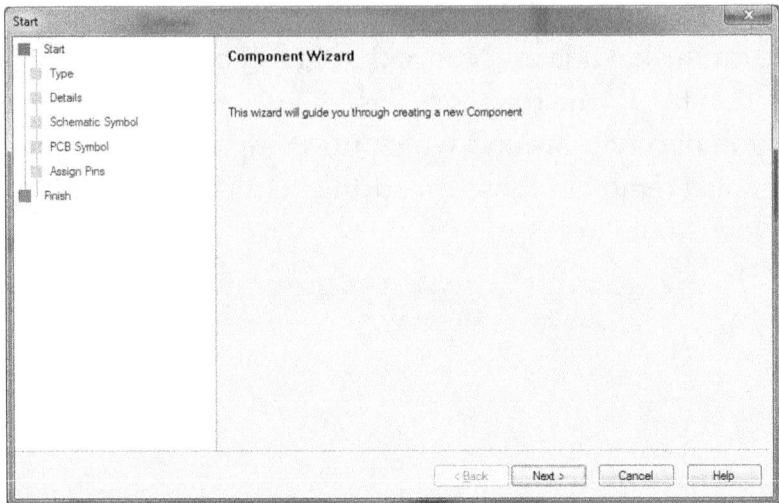

Type Next and select the normal Component.

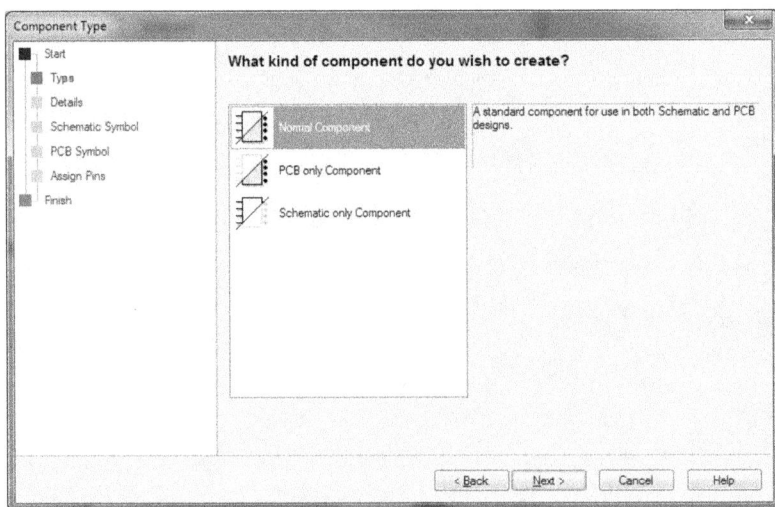

Type Next and type PotDualAudio into Component Name, SPICE into Package, P into Default Reference, 9 into Component Pins and 2 into number of gates.

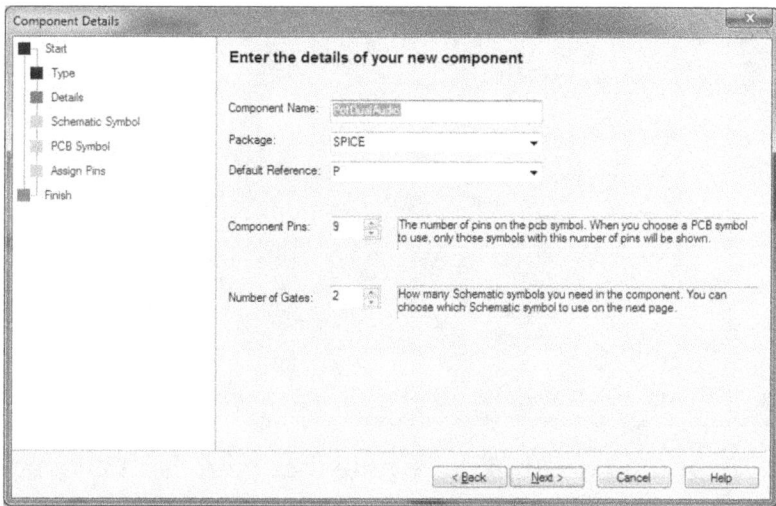

Select next and click on NewPot.

Select Next.

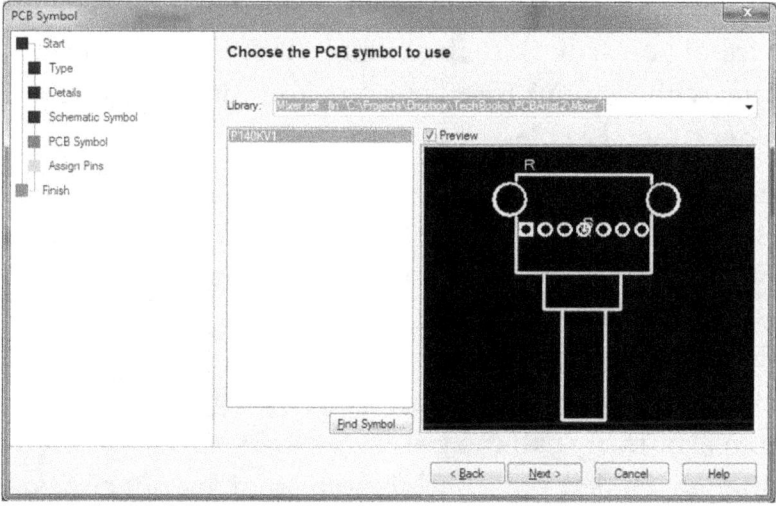

We can use our existing pot layout model so select P14KV1.

Select Next.

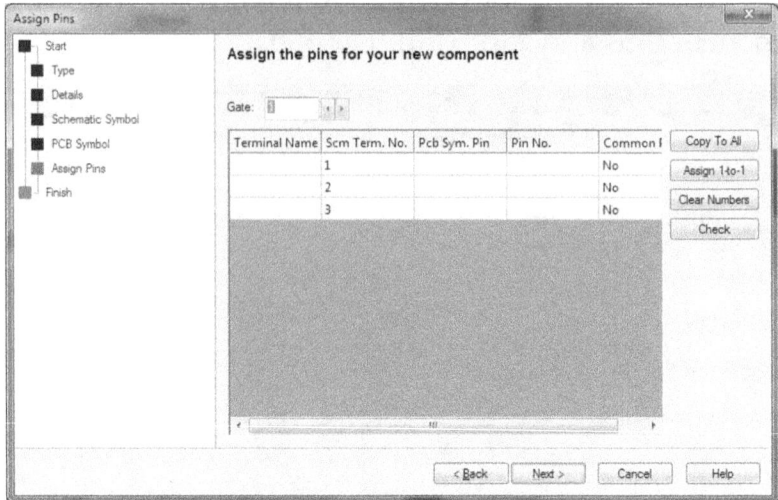

Before you do anything else note that there is a Gate: box near the top of the screen. Assign the pins for your new component section of the box.

When you first place the pot on the schematic you will place one half of the component. The next time you place the pot on the symbol you will place the other half of the symbol. This all happens without your intervention. You can change all these items once they are placed.

A glance back at the manufacturer's data shows that the pot's two sections are wired strangely. The wiring is excellent for manufacture but not so much so for layout. I am going to arbitrarily call the pot

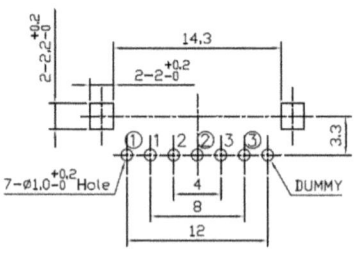

pins that are circled pot a and the pins that are not circled pot b.

Remember that the ScmTerm. No. corresponds to the

pattern in the model invocation.

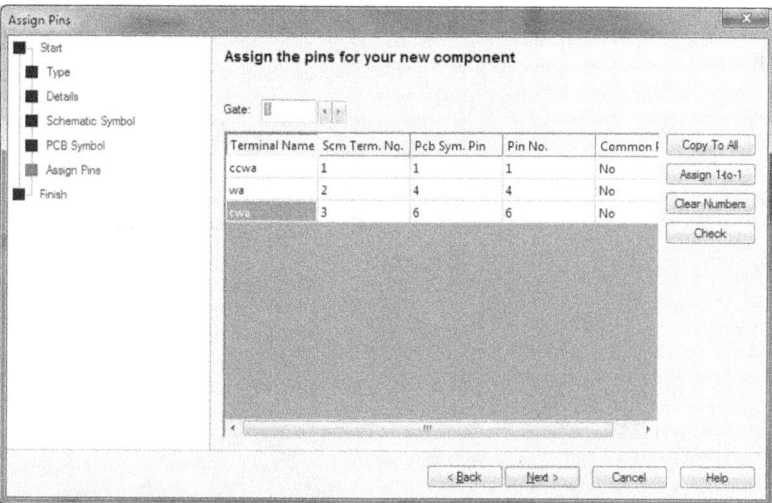

Then I pressed the right arrow in the Gate section selecting Gate b.

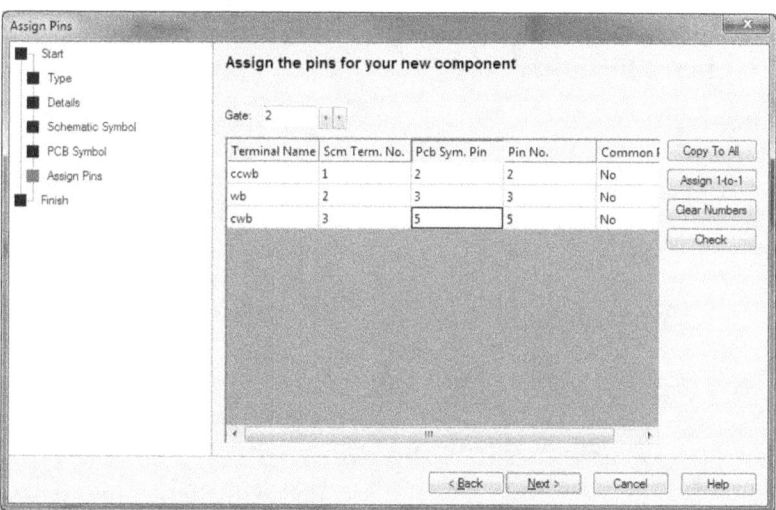

The pin names are ccw for Counter Clockwise with the wiper closest to this pin when the shaft is turned to the left. W stands for wiper and cw for Clockwise. The a and b suffixes are to denote which pot is being referenced.

Select Next.

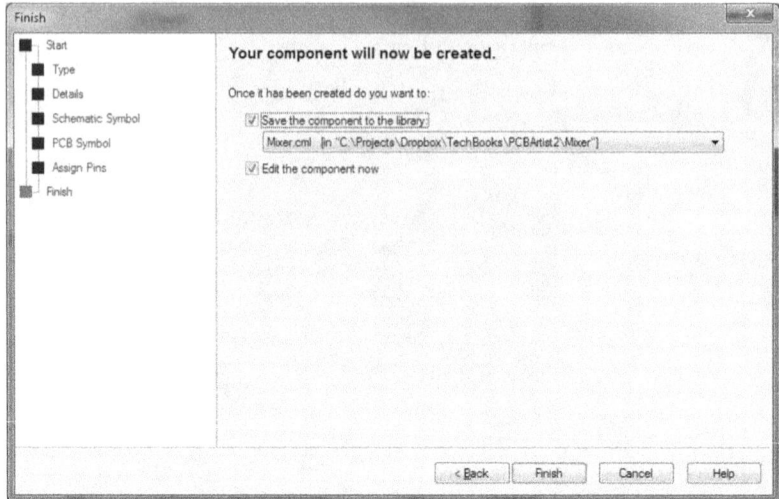

The schematic symbol is created.

Now that we have an appropriate schematic we can add the SPICE values. Type Cnt → L and select PotDualAudio form the Mixer.cml library.

Select Edit.

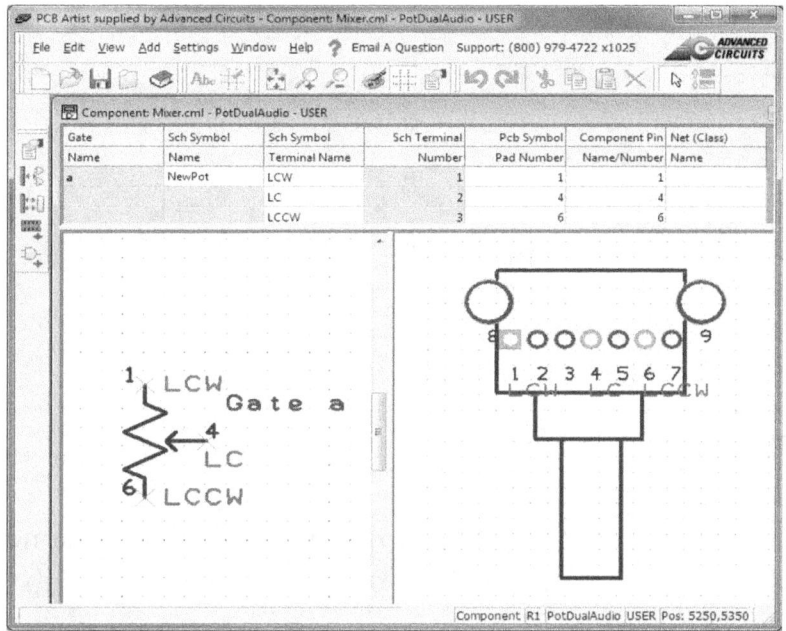

Adding the potentiometer to the circuit

This screen displays the model you just created and should be very close to the model we just created, especially the pin numbers. Spice uses the Sch Terminal Number to create the simulation models.

The model I'm going to use is complements of Helmut Sennewald and was obtained from:

http://tech.groups.yahoo.com/group/LTSpice/

This is a wonderful resource for LTSpice models and general information.

Type the following text into your favorite word processor and save as AudioPot.lib in the same directory as your Mixer schematic.

```
.SUBCKT pot_audio 1 2 3
* Parameters: Rtot, wiper, Rtap, Tap
.param w=limit(0.01m,wiper,0.99999)
*
.param pwrexp=ln(RTAP/RTOT)/(1-TAP)
.param ratio=exp(pwrexp*(1-w))
*
R1 1 2 {Rtot*(1-ratio)}
R2 2 3 {Rtot*(ratio)}
.ENDS
```

To use this model you need to know the .SUBCKT name which is pot_audio and include pins 1, 2 and 3 in your node list.

You will also need to provide the location of the model to LTSpice. This is accomplished by typing F8 and selecting Simulator Control in the LTSpice component library.

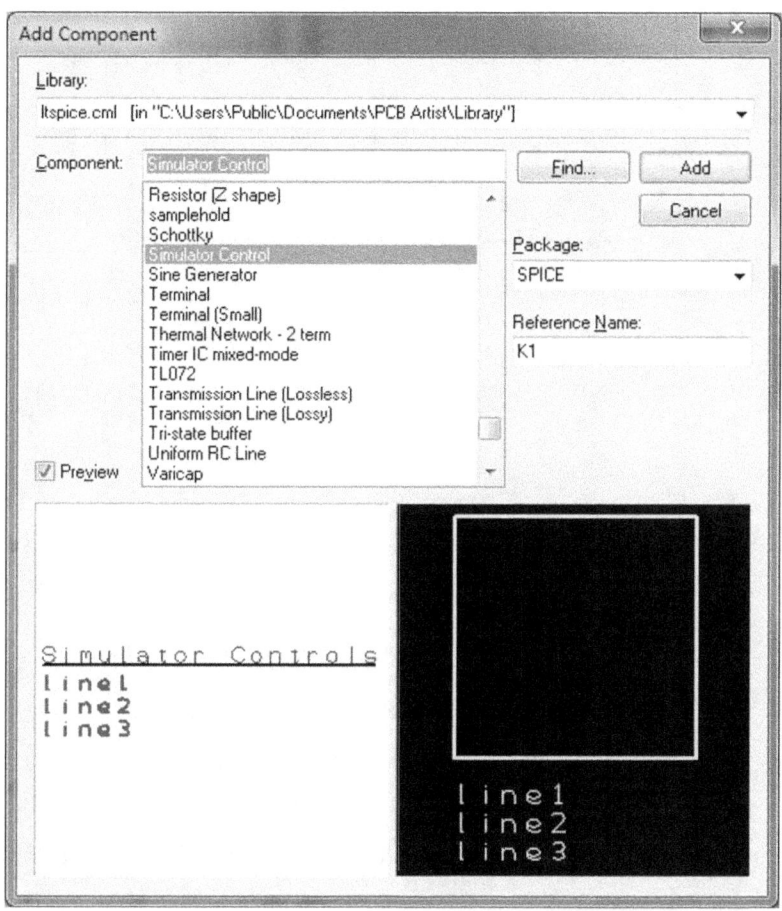

After you place the Simulator Control you right click it and select the Component Properties. Select Spice Template 1 and type in .lib AnalogPot.lib. If there are entries such as line 2 and line 3, for Spice Template 2 or Spice Template 3 you should erase those entries. If there are valid Simulator Controls (starting with a .) then you should keep them. You may have up to three Simulator Controls per symbol.

The checklist for adding this component is:

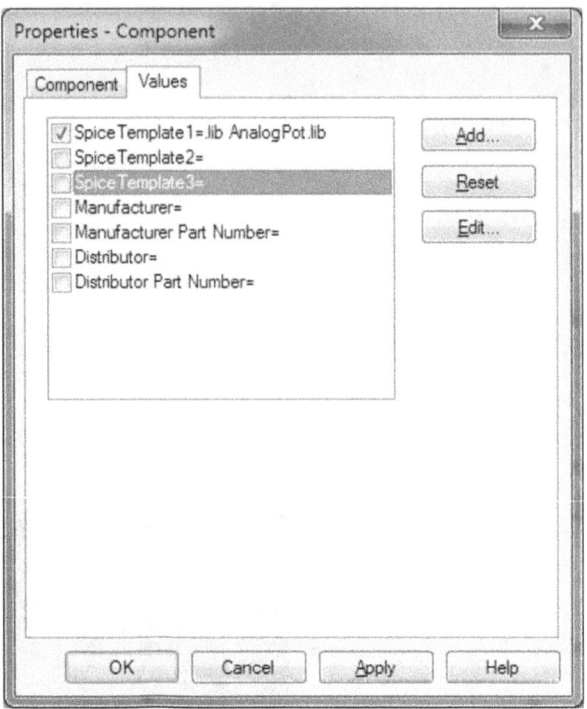

- Place AudioPot.lib into the local directory.
- Add the model we just created to the schematic.
- Add a simulator control to load AudioPot.lib into the simulator at run time.

Creating an Active Component Model

Active component models are somewhat more complicated. This is due to their non-linear nature. An active device, by definition, contains a non-linear element whether it be a diode, a transistor or even a tube. (A tube is a thermal device enveloped in a gas free environment.... never mind)

Component Type

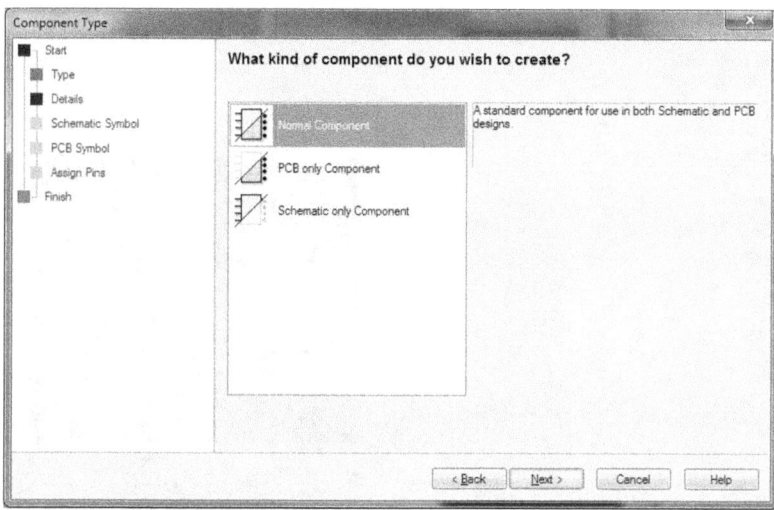

Component Details

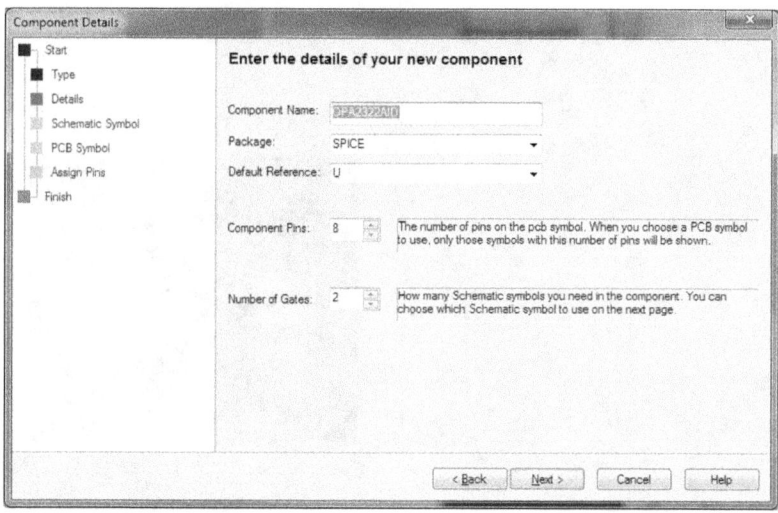

Schematic Symbol

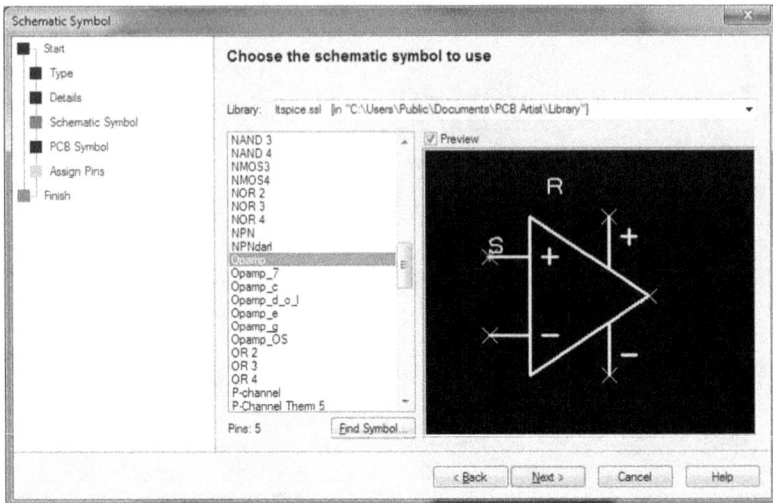

PCB Symbol

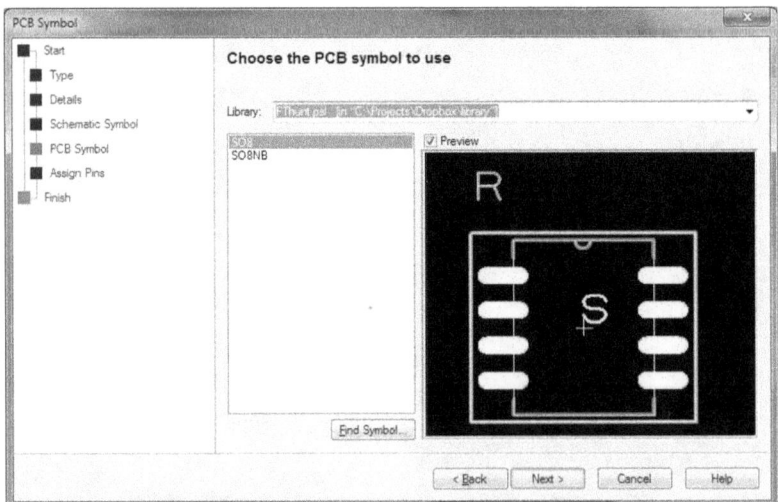

Assign Pins

Assigning pins is the most significant portion of adapting a model from another supplier. I have included the model for an OPAy322 (the y determines the number of opamps in a package – see the data sheet) You may notice that the model is from TI. The OPAy322 was selected for its characteristics instead of its convenience.

When you look at the model the line

`.SUBCKT OPA322 -IN +IN V- V+ Vout`

tells you pretty much everything you need to know about adapting the circuit. You MUST examine this line for every component that you adapt to LTSpice. Now lets look at the schematic model to which this model will be associated.

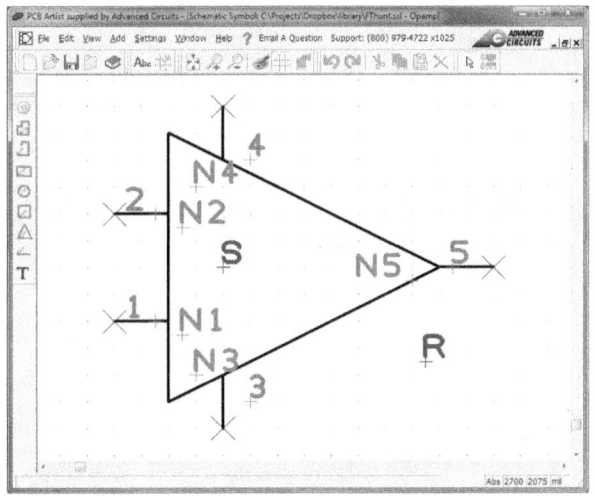

Notice that the pin number are 1 through 5 and bear no resemblance to the PCB symbol. The pin number 1 is associated with the first entry after the subcircuit name. Pin 1 is -IN, the second pin is +IN and so forth.

So when we get to the pin assignment it should look like:

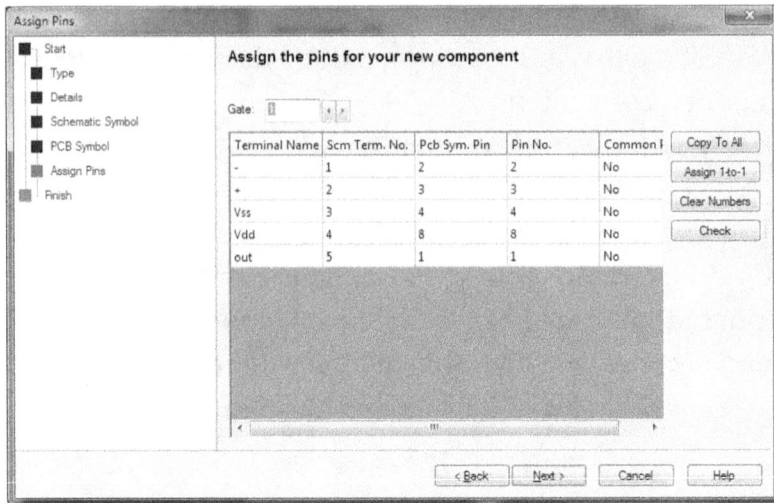

and

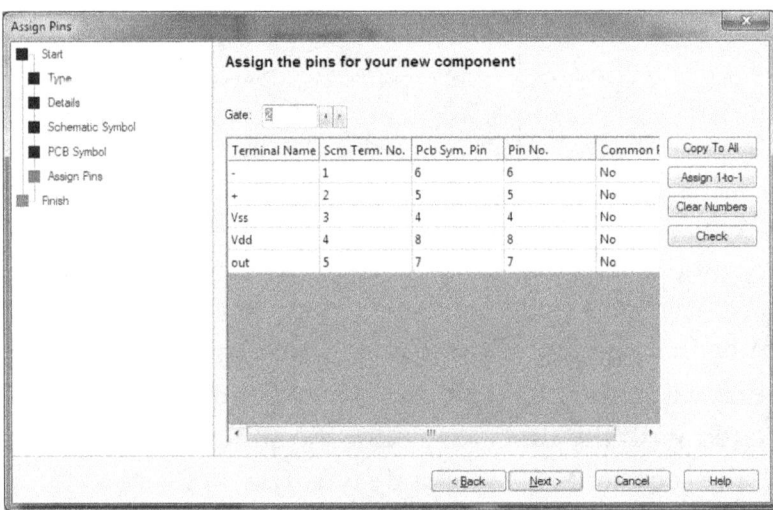

for Gates 1 and 2 respectively.

We now have a component symbol. I have included the library from Texas Instruments at the end of this chapter. Just a note, I have used TI's SPICE (TINA) more than LTSpice and find the product quite acceptable (and very similar.) I only selected LTSpice for the book because PCB Artist chose to make it their default simulator.

To use the OPAy322 model you must either type it in from this listing, or get the model from TI.com or get it in the download of the accompanying book software at www.AbrahamAaron.com .

If you type in the model save it as OPAy322.lib in project's local directory.

Next you have to add a line to the existing Simulator Controls window. We'll simply reuse the Simulator Controls window from the potentiometer section of this chapter.

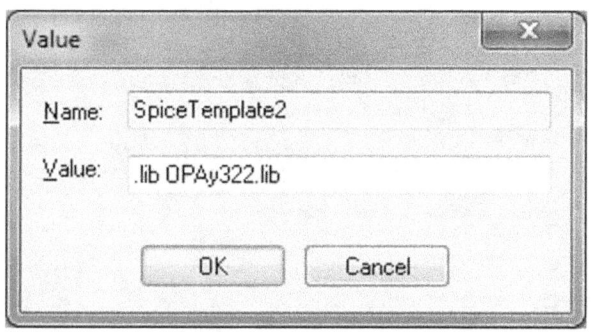

Add the information above and you should be able to stimulate through simulation.

The checklist is:

Create a schematic symbol

Find a model

Load the model via Simulator Controls.

The Hack

In the early 80's I was a member of the Tampa Bay Computer Club. At the time a hacker was a knowledgeable person that would find innovative methods to accomplish sometimes questionable goals with computer hardware. As usual the software guys abrogated the term and turned it into a term of disgust[12].

In our case we might have to simulate with a power source added and deleted during testing.....unless we have a dual purpose connector symbol. I've modified my connector to also be the power source so that I can simulate without having to add and remove a power source.

For this discussion I'll use the slang terms and terminology.

For the connector I've chosen pins 1 and 4 are the positive and negative voltages respectively. For the power supply model I modified the schematic pin numbers so that 1 and 2 match to the model.

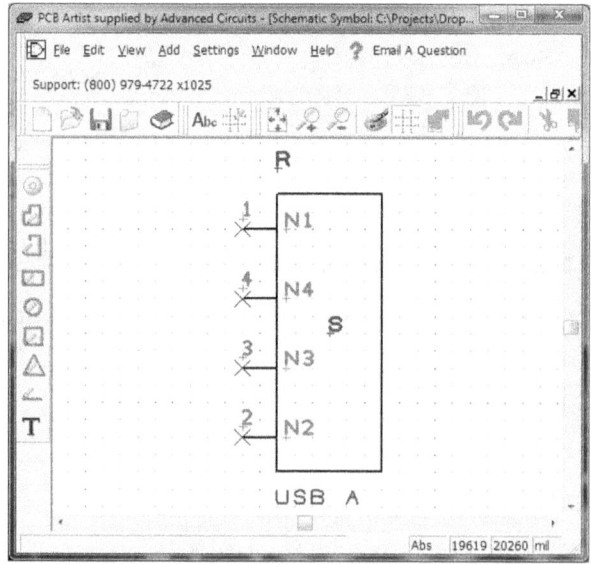

---

[12] Like the term active low for a tristate or totem pole output signal simply because the 'LOGIC' is true when low, the correct term. Active low means there is an active device for the purpose of sinking current to ground – and nothing else.

Then I mapped the schematic to the component like this.

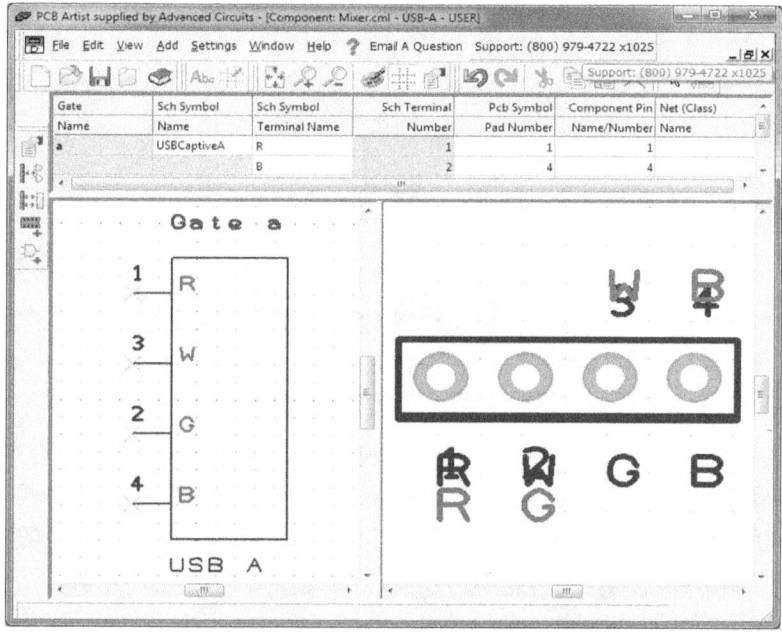

Then I hooked the Voltage Source model to my connector.

I'm sorry about the up and down and overwrite in the right side panel. I couldn't figure out how to fix it.

I added the line V and V=4.5 to the values for the connector and Voila, I have a connector that provides 4.5 Volts to power the circuit during simulation.

Again, PCB Artist and LTSpice have good documentation and their tutorials are excellent. The aim of this book is to demonstrate the use of those tools in a realistic (sort of) design environment.

The OPAy322.lib library

```
* Copyright 2010 by Texas Instruments Corporation

* GREEN-LIS MACRO-MODEL SIMULATED FEATURES:

* OPEN LOOP GAIN AND PHASE VS FREQUENCY WITH RL AND CL EFFECTS
* INPUT COMMON MODE REJECTION WITH FREQUENCY
* POWER SUPPLY REJECTION WITH FREQUENCY
* INPUT IMPEDANCE VS FREQUENCY
* OUTPUT IMPEDANCE VS FREQUENCY FOR DIFFERENT OUTPUT CURRENTS
* INPUT VOLTAGE NOISE VS FREQUENCY
* INPUT CURRENT NOISE VS FREQUENCY
* OUTPUT VOLTAGE SWING VS OUTPUT CURRENT
* SHORT-CIRCUIT OUTPUT CURRENT
* QUIESCENT CURRENT VS SUPPLY VOLTAGE
* SETTLING TIME VS CAPACITIVE LOAD
* SLEW RATE
* SMALL SIGNAL OVERSHOOT VS CAPACITIVE LOAD
* LARGE SIGNAL RESPONSE
* OVERLOAD RECOVERY TIME
* INPUT BIAS CURRENT
* INPUT VOLTAGE OFFSET
* INPUT COMMON MODE RANGE
* OUTPUT CURRENT COMING THROUGH THE SUPPLY RAILS

.SUBCKT OPA322 -IN +IN V- V+ Vout
V7          48 12 0
Vos         28 39 -445U
V11         53 54 100M
V10         55 56 100M
V6          11 61 10
V5          62 11 10
V4          58 60 1
V1          59 57 1
V9          73 13 0
IS2         V+ 28 400F
IS1         V+ V- 1.5M
IS3         44 V- -200F
V3          77 11 63
V2          11 78 73
XU15        14 11 29 30 VC_RES_0
L1          31 11 750P IC=0
R2          31 32 1
GVCCS8      11 32 11 33  1
XR109       34 11 RNOISE_FREE_0
C3          34 11 5F
GVCCS4      11 34 25 11  1M
```

```
C2              35 11 5F
XR109_2         35 11 RNOISE_FREE_0
GVCCS3          11 35 34 11  1M
R4              36 22 10M
C7              37 38 3P
C8              38 11 4P
CinnCM          11 37 4P
XIn11           39 37 FEMT_0
L2              40 11 2U IC=0
XR109_3         25 11 RNOISE_FREE_1
XR109_4         41 11 RNOISE_FREE_1
XVn11           38 39 VNSE_0
XU14            42 11 43 44 VCVS_LIMIT_0
L3              45 11 1U IC=0
R1              40 42 1
GVCCS2          11 42 11 46  1M
XU13            12 47 IDEAL_D_0
EVCVS5          48 11 V- 11  1
XR109_5         49 11 RNOISE_FREE_1
C11             41 11 2.5F
XR109_6         24 11 RNOISE_FREE_0
GVCCS12         11 25 41 11  1U
XU5             14 11 V+ 15 VCVS_LIMIT_1
XU6             11 14 16 V- VCVS_LIMIT_2
C15             V+ V- 10P IC=0
C22             11 21 1P
R29             21 23 1
C23             11 26 1P IC=0
C9              50 11 10P IC=0
R26             50 14 10
C21             11 17 1P IC=0
C20             11 18 1P IC=0
C19             19 11 1P IC=0
C17             20 11 1P IC=0
C16             11 51 1P IC=0
C12             52 11 1P IC=0
R13             30 26 1
R36             26 56 1M
R35             26 54 1M
SW12            57 53 19 11   S_VSWITCH_1
SW11            55 58 11 20   S_VSWITCH_2
R34             26 59 1K
R33             26 60 1K
SW10            62 23 21 11   S_VSWITCH_3
SW9             23 61 11 21   S_VSWITCH_4
R25             63 19 1
R19             64 20 1
R16             65 51 1
R14             66 52 1
R12             67 17 1
```

```
R7              68 18 1
R5              69 24 10M
R6              70 23 10M
R15             0 11 100MEG
C13             24 11 5F
GVCCS1          11 24 35 11   1M
GIsinking       V- 11 71 11   1M
GIsourcing      V+ 11 72 11   1M
R23             71 11 10K
SW7             14 71 50 11   S_VSWITCH_5
R21             11 72 10K
SW8             14 72 50 11   S_VSWITCH_6
SW4             70 67 17 11   S_VSWITCH_7
SW3             68 70 11 18   S_VSWITCH_8
XU3             58 27 68 11 VCVS_LIMIT_3
XU1             57 27 67 11 VCVS_LIMIT_3
SW2             36 63 19 11   S_VSWITCH_9
SW1             64 36 11 20   S_VSWITCH_10
XU8             28 V+ IDEAL_D_1
XU12            V- 28 IDEAL_D_1
EVCVS6          73 11 V+ 11   1
R22             74 47 100
EVCVS4          74 11 28 11   1
XU2             47 13 IDEAL_D_0
SW6             69 65 51 11   S_VSWITCH_11
SW5             66 69 11 52   S_VSWITCH_12
XU26            47 44 11 75 VCCS_LIMIT_0
XU4             75 11 11 23 VCCS_LIMIT_1
LPSR            76 11 750U IC=0
XVCVSPSRR       32 11 43 37 VCVS_LIMIT_4
XU22            77 14 64 11 VCVS_LIMIT_5
XU21            78 14 63 11 VCVS_LIMIT_5
XU20            16 Vout 65 11 VCVS_LIMIT_5
XU19            15 Vout 66 11 VCVS_LIMIT_6
XU11            V- 44 IDEAL_D_1
XU10            44 V+ IDEAL_D_1
C10             22 11 5F
C5              25 11 2.5F
XR109_7         22 11 RNOISE_FREE_0
GVCCS15         11 22 24 11   1M
GVCCS10         11 41 49 11   1U
R20             +IN 38 100
R18             -IN 37 100
GVCCS6          11 49 27 11   1U
XR102           79 80 RNOISE_FREE_1
XR101           81 79 RNOISE_FREE_1
C6              79 0 1 IC=0
XR105           27 11 RNOISE_FREE_1
XR104           23 11 RNOISE_FREE_2
XR103           11 75 RNOISE_FREE_1
```

```
EVCVS34        11 0 79 0  1
RPSR           76 33 1
GVCCS11        11 33 V+ V-   10U
RCM            45 46 1
EVCVS29        81 0 V+ 0  1
EVCVS28        80 0 V- 0  1
GVCCS7         11 46 28 11  10M
VCCVS1_in      29 Vout
HCCVS1         14 11 VCCVS1_in   1K
GVCCS5         11 27 23 11  1U
Ccc            23 11 700N
EVCVS3         30 11 22 11  1
.MODEL S_VSWITCH_1 VSWITCH (RON=1 ROFF=10MEG VON=100M VOFF=-100M)
.MODEL S_VSWITCH_2 VSWITCH (RON=1 ROFF=10MEG VON=100M VOFF=-100M)
.MODEL S_VSWITCH_3 VSWITCH (RON=10M ROFF=100MEG VON=50 VOFF=30)
.MODEL S_VSWITCH_4 VSWITCH (RON=10M ROFF=100MEG VON=50 VOFF=30)
.MODEL S_VSWITCH_5 VSWITCH (RON=1M ROFF=10MEG VON=-10M VOFF=0)
.MODEL S_VSWITCH_6 VSWITCH (RON=1M ROFF=10MEG VON=10M VOFF=0)
.MODEL S_VSWITCH_7 VSWITCH (RON=1 ROFF=10MEG VON=1 VOFF=-1)
.MODEL S_VSWITCH_8 VSWITCH (RON=1 ROFF=10MEG VON=1 VOFF=-1)
.MODEL S_VSWITCH_9 VSWITCH (RON=1 ROFF=1G VON=10 VOFF=-10)
.MODEL S_VSWITCH_10 VSWITCH (RON=1 ROFF=1G VON=10 VOFF=-10)
.MODEL S_VSWITCH_11 VSWITCH (RON=1 ROFF=1G VON=10 VOFF=-10)
.MODEL S_VSWITCH_12 VSWITCH (RON=1 ROFF=1G VON=10 VOFF=-10)
.ENDS

*VOLTAGE CONTROLLED RESISTOR
.SUBCKT VC_RES_0  1       2       3     4
*                 VC+     VC-    RES1 RES2
ERES 3 40 VALUE = {92*(I(VSENSE)/SQRT((.5*ABS(V(1,2))+0.8)/0.8))}
VSENSE 40 4 DC 0
.ENDS VC_RES_0

* NOISELESS RESISTOR
.SUBCKT RNOISE_FREE_0  1 2
*ROHMS = VALUE IN OHMS OF NOISELESS RESISTOR
.PARAM ROHMS=1E3
ERES 1 3 VALUE = { I(VSENSE) * ROHMS }
RDUMMY 30 3 1
VSENSE 30 2 DC 0V
.ENDS RNOISE_FREE_0

* BEGIN PROG NSE FEMTO AMP/RT-HZ
.SUBCKT FEMT_0  1 2
* BEGIN SETUP OF NOISE GEN - FEMPTOAMPS/RT-HZ
* INPUT THREE VARIABLES
* SET UP INSE 1/F
* FA/RHZ AT 1/F FREQ
.PARAM NLFF=.1
* FREQ FOR 1/F VAL
```

```
.PARAM FLWF=0.001
* SET UP INSE FB
* FA/RHZ FLATBAND
.PARAM NVRF=.1
* END USER INPUT
* START CALC VALS
.PARAM GLFF={PWR(FLWF,0.25)*NLFF/1164}
.PARAM RNVF={1.184*PWR(NVRF,2)}
.MODEL DVNF D KF={PWR(FLWF,0.5)/1E11} IS=1.0E-16
* END CALC VALS
I1 0 7 10E-3
I2 0 8 10E-3
D1 7 0 DVNF
D2 8 0 DVNF
E1 3 6 7 8 {GLFF}
R1 3 0 1E9
R2 3 0 1E9
R3 3 6 1E9
E2 6 4 5 0 10
R4 5 0 {RNVF}
R5 5 0 {RNVF}
R6 3 4 1E9
R7 4 0 1E9
G1 1 2 3 4 1E-6
C1 1 0 1E-15
C2 2 0 1E-15
C3 1 2 1E-15
.ENDS
* END PROG NSE FEMTO AMP/RT-HZ


* NOISELESS RESISTOR
.SUBCKT RNOISE_FREE_1  1 2
*ROHMS = VALUE IN OHMS OF NOISELESS RESISTOR
.PARAM ROHMS=1E6
ERES 1 3 VALUE = { I(VSENSE) * ROHMS }
RDUMMY 30 3 1
VSENSE 30 2 DC 0V
.ENDS RNOISE_FREE_1


* BEGIN PROG NSE NANO VOLT/RT-HZ
.SUBCKT VNSE_0  1 2
* BEGIN SETUP OF NOISE GEN - NANOVOLT/RT-HZ
* INPUT THREE VARIABLES
* SET UP VNSE 1/F
* NV/RHZ AT 1/F FREQ
.PARAM NLF=56
* FREQ FOR 1/F VAL
.PARAM FLW=10
* SET UP VNSE FB
```

```
* NV/RHZ FLATBAND
.PARAM NVR=6.6
* END USER INPUT
* START CALC VALS
.PARAM GLF={PWR(FLW,0.25)*NLF/1164}
.PARAM RNV={1.184*PWR(NVR,2)}
.MODEL DVN D KF={PWR(FLW,0.5)/1E11} IS=1.0E-16
* END CALC VALS
I1 0 7 10E-3
I2 0 8 10E-3
D1 7 0 DVN
D2 8 0 DVN
E1 3 6 7 8 {GLF}
R1 3 0 1E9
R2 3 0 1E9
R3 3 6 1E9
E2 6 4 5 0 10
R4 5 0 {RNV}
R5 5 0 {RNV}
R6 3 4 1E9
R7 4 0 1E9
E3 1 2 3 4 1
C1 1 0 1E-15
C2 2 0 1E-15
C3 1 2 1E-15
.ENDS

    END PROG NSE NANOV/RT-HZ

 •
*VOLTAGE CONTROLLED SOURCE WITH LIMITS
.SUBCKT VCVS_LIMIT_0  VC+ VC- VOUT+ VOUT-
*
.PARAM GAIN = 1
.PARAM VPOS = 10M
.PARAM VNEG = -10M
E1 VOUT+ VOUT- VALUE={LIMIT(GAIN*V(VC+,VC-),VPOS,VNEG)}
.ENDS VCVS_LIMIT_0

*TG IDEAL DIODE
.SUBCKT IDEAL_D_0  A C
D1 A C DNOM
.MODEL DNOM D (TT=10P CJO=1E-18 IS=1E-15 RS=1E-3)
.ENDS IDEAL_D_0

*VOLTAGE CONTROLLED SOURCE WITH LIMITS
.SUBCKT VCVS_LIMIT_1  VC+ VC- VOUT+ VOUT-
*
```

```
   E1 VOUT+ VOUT- TABLE {ABS(V(VC+,VC-))} = (0,0.12) (6,0.12)
(20,0.32) (30,0.53) (40,0.8) (50,1.17) (58,1.8) (62.9,2.7)
   .ENDS VCVS_LIMIT_1


   *VOLTAGE CONTROLLED SOURCE WITH LIMITS
   .SUBCKT VCVS_LIMIT_2  VC+ VC- VOUT+ VOUT-
   *

   E1 VOUT+ VOUT- TABLE {ABS(V(VC+,VC-))} = (0,0.12) (8,0.12)
(20,0.3) (30,0.48) (40,0.68) (50,0.9) (60, 1.2) (70,2.0) (72.9,2.7)
   .ENDS VCVS_LIMIT_2

   *VOLTAGE CONTROLLED SOURCE WITH LIMITS
   .SUBCKT VCVS_LIMIT_3  VC+ VC- VOUT+ VOUT-
   *
   .PARAM GAIN = 100
   .PARAM VPOS = 6000
   .PARAM VNEG = -6000
   E1 VOUT+ VOUT- VALUE={LIMIT(GAIN*V(VC+,VC-),VPOS,VNEG)}
   .ENDS VCVS_LIMIT_3

   *TG IDEAL DIODE
   .SUBCKT IDEAL_D_1  A C
   D1 A C DNOM
   .MODEL DNOM D (TT=10P CJO=1E-18 IS=1E-15 RS=1E-3)
   .ENDS IDEAL_D_1

   *VOLTAGE CONTROLLED SOURCE WITH LIMITS
   .SUBCKT VCCS_LIMIT_0  VC+ VC- IOUT+ IOUT-
   *
   .PARAM GAIN = 100U
   .PARAM IPOS = .5
   .PARAM INEG = -.5
   G1 IOUT+ IOUT- VALUE={LIMIT(GAIN*V(VC+,VC-),INEG,IPOS)}
   .ENDS VCCS_LIMIT_0

   *VOLTAGE CONTROLLED SOURCE WITH LIMITS
   .SUBCKT VCCS_LIMIT_1  VC+ VC- IOUT+ IOUT-
   *
   .PARAM GAIN = 1
   .PARAM IPOS = 7
   .PARAM INEG = -7
   G1 IOUT+ IOUT- VALUE={LIMIT(GAIN*V(VC+,VC-),IPOS,INEG)}
   .ENDS VCCS_LIMIT_1

   *VOLTAGE CONTROLLED SOURCE WITH LIMITS
   .SUBCKT VCVS_LIMIT_4  VC+ VC- VOUT+ VOUT-
   *
   .PARAM GAIN = -1
   .PARAM VPOS = 10M
```

```
.PARAM VNEG = -10M
E1 VOUT+ VOUT- VALUE={LIMIT(GAIN*V(VC+,VC-),VPOS,VNEG)}
.ENDS VCVS_LIMIT_4

*VOLTAGE CONTROLLED SOURCE WITH LIMITS
.SUBCKT VCVS_LIMIT_5  VC+ VC- VOUT+ VOUT-
*
.PARAM GAIN = 100
.PARAM VPOS = 5000
.PARAM VNEG = -5000
E1 VOUT+ VOUT- VALUE={LIMIT(GAIN*V(VC+,VC-),VPOS,VNEG)}
.ENDS VCVS_LIMIT_5

*VOLTAGE CONTROLLED SOURCE WITH LIMITS
.SUBCKT VCVS_LIMIT_6  VC+ VC- VOUT+ VOUT-
*
.PARAM GAIN = 100
.PARAM VPOS = 5000
.PARAM VNEG = -5000
E1 VOUT+ VOUT- VALUE={LIMIT(GAIN*V(VC+,VC-),VPOS,VNEG)}
.ENDS VCVS_LIMIT_6

* NOISELESS RESISTOR
.SUBCKT RNOISE FREE_2  1 ?
*ROHMS = VALUE IN OHMS OF NOISELESS RESISTOR
.PARAM ROHMS=1E4
ERES 1 3 VALUE = { I(VSENSE) * ROHMS }
RDUMMY 30 3 1
VSENSE 30 2 DC 0V
.ENDS RNOISE_FREE_2


.END
```

Spice Suffix Nomenclature

Letter suffix	Word Suffix	Magnitude	English
T	Terra	10^{12}	trillion
G	Giga	10^{9}	billion
Meg	Mega	10^{6}	million
K	Kilo	10^{3}	thousand
M	Milli	10^{-3}	thousandth
U	Micro	10^{-6}	Millionth
N	Nano	10^{-9}	Billionth
P	Pico	10^{-12}	Trillionth
F	Femto	10^{-15}	Oh-come-on

Do be careful and not use M for Mega as this will be interpreted as 1/1000 th.

Typical schematic symbols

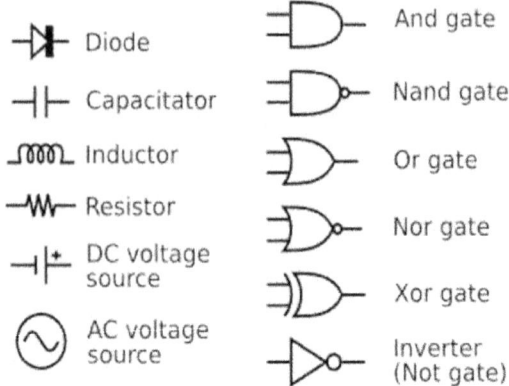

⊶▶⊢ Diode	And gate
⊣⊢ Capacitor	Nand gate
⌁ Inductor	Or gate
⌁ Resistor	Nor gate
⊣⊢ DC voltage source	Xor gate
⊘ AC voltage source	Inverter (Not gate)

Audio Mixer Product Design Requirements

The labels before each section below are redundant, but included to provide a methodology for stating requirements.

NEED: The general concept is to mix multiple electronic sound sources into a single signal for a listener's direct enjoyment or for amplification. The specific intention is to provide an interface between multiple sound sources such as a personal computer sound output and a secure MP3 player. The secure MP3 player might be enabled to play music that would not be on the computer if, for example, the computer belongs to another entity or the content can only be played from one source or, as in the case of an Audible book, the MP3 player allows the listener to keep their place in the file while in transit between home and the office computer.

BACKGROUND: While not yet the dominate method of connection between sound sources, the stereo 3.5mm connector is rapidly becoming the de facto standard for connecting consumer sound recording and playback devices. Examples are headphones, computer sound ports and television audio.

Sound sources provide outputs of different levels and some are either not adjustable or must be adjusted to provide for other needs. The solution is to provide volume controls to adjust the channels.

Most of the devices above will have standard USB ports. The standard USB port can provide inexpensive and convenient power.

NAME: The product is an audio mixer or mixer.

PURPOSE: The mixer combines a three audio inputs into a single audio input.

DESCRIPTION: Three 3.5 mm female input sockets are provided that accept standard line level 0-1V stereo outputs from audio devices such as MP3 players, computers, tablets, cell phones and similar devices. There is one input level control for each stereo channel. The input signals are combined (mixed) into a single stereo output signal. The output connector is a 3.5mm female output socket. There is a single output level control.

POWER: Power for the mixer is by means of a standard USB Type B connector.

ENCLOSURE: The unit will fit into a Hammond 1591XXATBU enclosure. DK part number HM1051-ND.

SO-8 Package

Example Board Layout
(Note C)

Stencil Openings
(Note D)

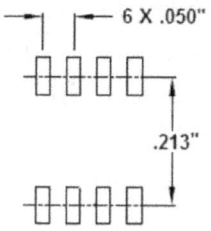

6 X .050"

.213"

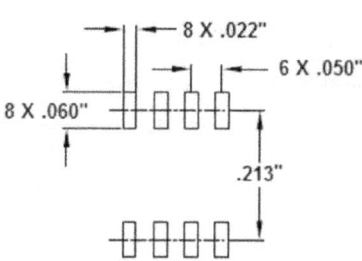

8 X .022"

6 X .050"

8 X .060"

.213"

P140KV1 Through Hole Package

We are using the P140KV1-F25BR10K potentiometer from TT Electronics. This is a dual 10K linear potentiometer. A copy of the mechanical portion of the data sheet is included. The measurements have been changed from mm to inches.

Model P140KV1 (Side Adjust, with Bushing)

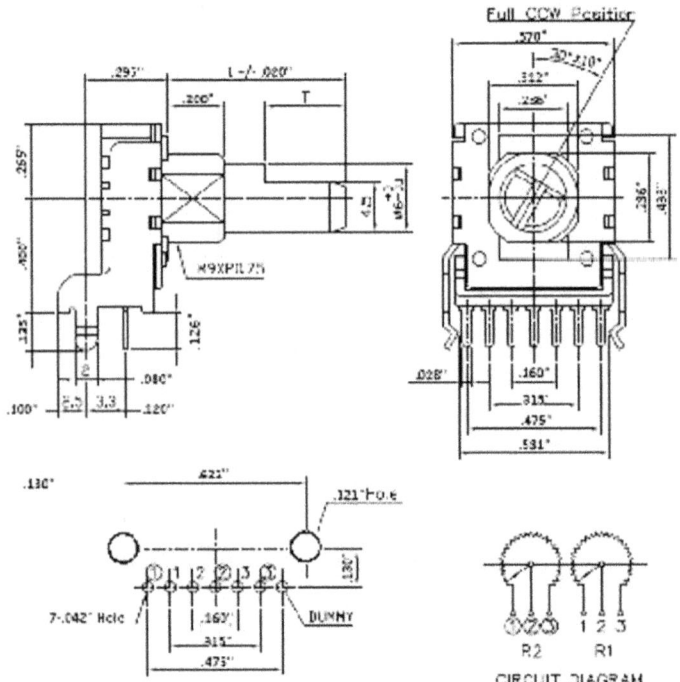

Schematic Drawings

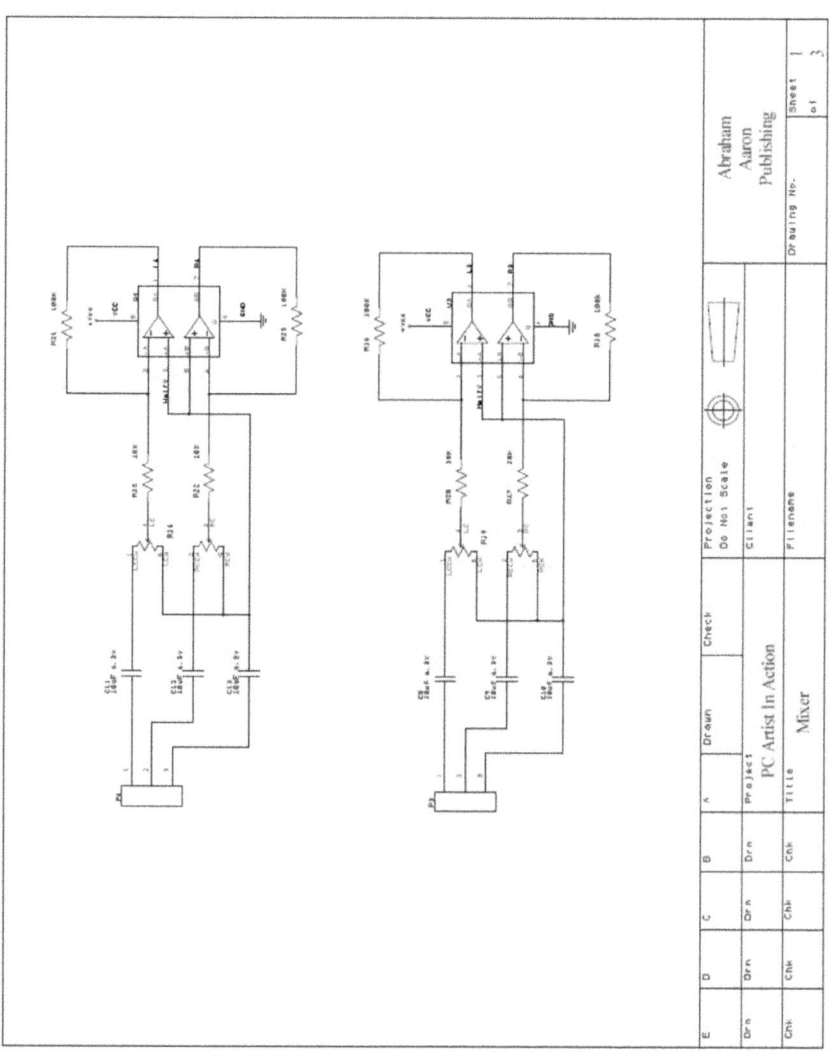

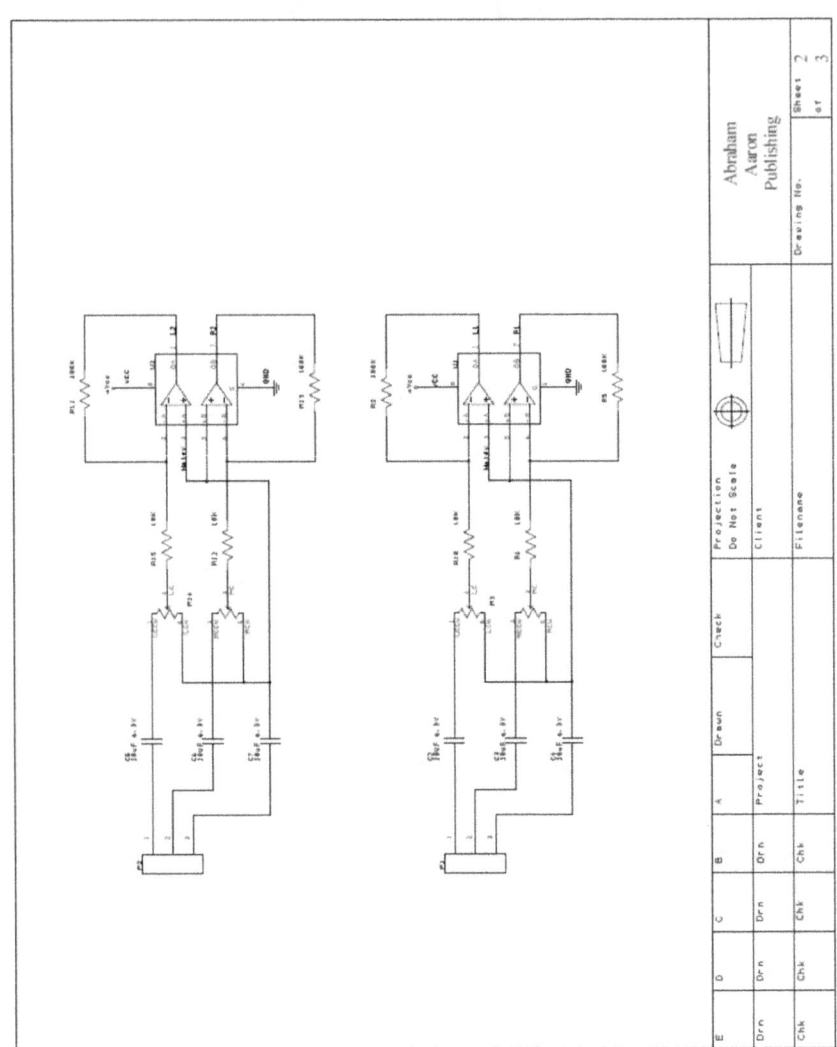

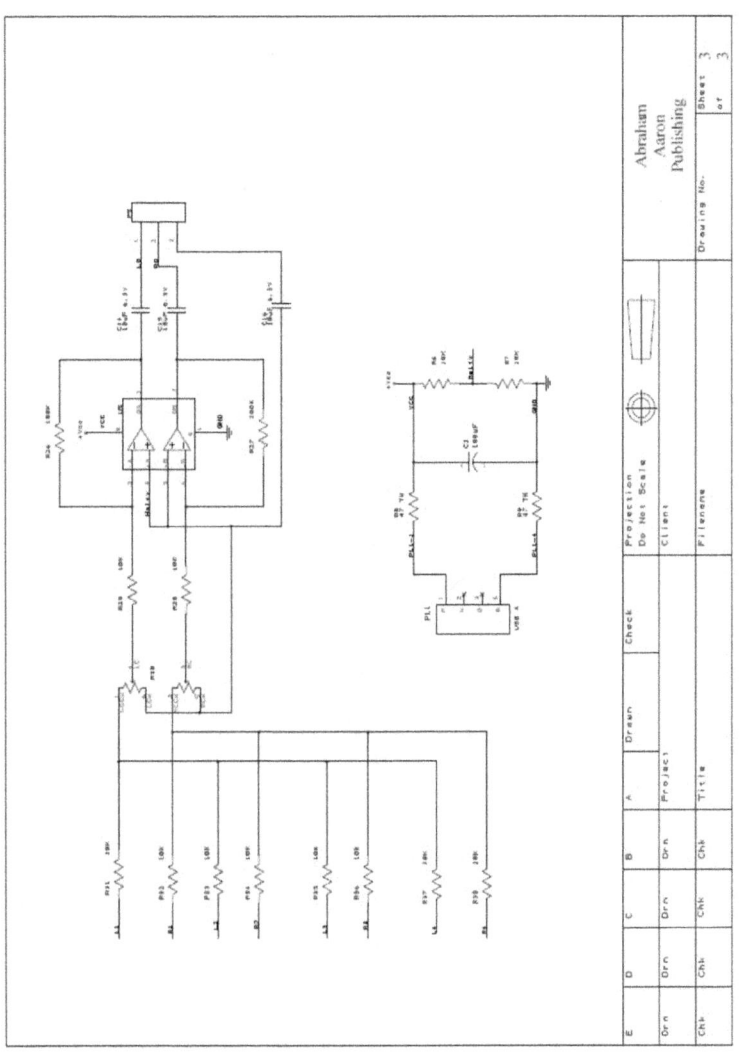

www.ingramcontent.com/pod-product-compliance
Lightning Source LLC
Chambersburg PA
CBHW051447170526
45166CB00001B/148